Table

Key: The bullets used below and i

- Featured rock (includes rock ic
- Rock or mineral is described bu
- Mineral (not a rock).

* Varieties of these rocks and minerals from the Great Lakes region are included as examples.

Introduction..1
Geologic Background: A Few Basic Facts................4
Rock Identification: The Starting Point.................7

Chapter 1: Source of Minerals..............................9

Chapter 2: Intrusive Igneous Rocks
Geologic Background...15
Featured Rocks..21
- Granite...22
- Pegmatite..32
- Granodiorite...38
- Diorite...39
- Gabbro..46
- Syenite..54

Chapter 3: Extrusive Igneous Rocks
Geologic Background...60
Featured Rocks..67
- Basalt..68
  - Vesicular Basalt
  - Scoria
  - Porphyritic Basalt
  - Amygdaloidal Basalt
- Andesite..82
- Rhyolite..85
  - Pumice

iii

Chapter 3: Extrusive Igneous Rocks (continued)

- **Amygdaloidal Minerals**...93
    - Chalcedony...93
    - Agate..97
    - Carnelian..104
    - Calcite..107
    - Chlorastrolite (Isle Royale Greenstone)*.........108
    - Copper...109
    - Datolite*..112
    - Epidote...113
    - Milky quartz..115
    - Prehnite*...118
    - Thomsonite*..119

Chapter 4: Sedimentary Rocks

Geologic Background..120
Featured Rocks...123
- **Banded Iron Formation**..................................124
- **Breccia**...130
- **Chert**..137
    - Jasper..143
- **Conglomerate**...147
    - Copper Harbor Conglomerate*....................148
    - Puddingstone...150
- **Sandstone**..154
    - Graywacke
        - Omarolluk (Omar)*
    - Jacobsville Sandstone*
    - Munising Formation Sandstone*
- **Limestone**..172
- **Shale**...179

iv

Rocks
Inside Out

Designed, written, and compiled by

Karen Brzys
www.agatelady.com

Rocks Inside Out

Copyright ©2021 by Karen Brzys (Gitche Gumee Agate and History Museum).

Printed by Versa Press. First printing June 2021

ISBN-13: 978-0-9760559-5-2

All rights are reserved. No part of this book may be reproduced or copied by any means including electronic, digital, photocopying, mechanical, or any other method without written permission from Karen Brzys.

For information about obtaining permission, or to purchase additional copies, please contact: Karen Brzys, PO Box 308, Grand Marais, MI 49839; www.agatelady.com; Karen@agatelady.com. If you have any questions or comments, or if you find any typographical, grammatical, or other errors in this book, please send an email to Karen@agatelady.com. Thanks in advance for your feedback.

Dedications and Acknowledgements

This book is dedicated to the two people most responsible for providing me with the skills and knowledge required to complete this project. Axel Niemi, my mentor and the founder of the Gitche Gumee Museum, sparked my interest when I was a kid and taught me much about the fascinating world of rocks, minerals, and geology. Nadine Dyer, my PhD high school English teacher, taught me how to write. My life would have certainly been much different without these two incredible individuals. I would also like to dedicate this book to the many thousands of people who have visited either the museum or my booths at rock shows. Without the financial support you provided with your gift shop and rock show purchases, the opportunity for me to write books would not have been possible.

The effort to create this book was significantly bolstered by others. First in line to be acknowledged are the geology reviewers, Dr. William Cordua and James St. John. Both earned additional recognition for their high resolution photographs that made their way into this book, plus their editing that went beyond the call of duty. Senior editor, Briana Rupel contributed her grammatical and basic editing skills. Other reviewers who proofread the manuscript included Renee Beaver, Karen Boaz, Marsha Hendrickson, Diana Mavis, Doug Moore, Jill and Gerald Phillips, Helen Riley, and Claudia Wyrick. Finally, thanks to all my friends and family who are patient with me when I am singly focused and working on books. One of my friends suggests that when a book is in progress, I am like a dog with a new bone. I have deep gratitude for everyone who has provided support and assistance. Thank you from the bottom of my heart.

Chapter 5: Metamorphic Rocks
    Geologic Background..185
    Featured Rocks...189
        - Slate..189
        - Schist..195
        - Gneiss...202
        - Unakite...208

Appendices
    A. Instructions for Testing the Hardness of Rocks and Minerals..........213
    B. Rock Cycle Diagram..215
    C. Quick Reference Identification Chart...216
    D. Geologic Timeline...218
    E. Bibliography..220
    F. Figure Attributions..223
    G. Index..232

NOTE: As much as possible, this book has been written without jargonistic language. However, to effectively discuss geology, describe how rocks formed, and explain the characteristics needed to identify rocks – some terminology is required. In many cases, the terms are defined when they are used, especially the first time the term appears in the book. Most of the terms are included in the index, so all the pages with that term are listed. Since it is easier these days to do an Internet search for a definition than it is to look up terms in a glossary, a glossary has not been included.

Rock Hunting Etiquette and Advice

1. Please respect private property and ask permission before entering.
2. Research laws and regulations governing rock hunting in the target search area.
3. It is not legal to remove rocks and minerals from national parks and national lakeshores.
4. While searching for geologic treasures, be respectful of the environment and leave no trace.
5. Support the Rockhound Project H.E.L.P. (Help Eliminate Litter Please). Bring a garbage bag and pick up other people's trash.
6. Do not hoard. Leave rocks for others, especially rare varieties.
7. Limit excavation depth to four feet (1.2 m). Fill in all search holes before leaving.
8. If you enter through gates to legally access a rock hunting area, leave all gates as found (either open or closed).
9. Research weather conditions before heading out.
10. Prior to departure, let someone know where you are going and when you expect to return. Let your contact know when you are safely home.
11. If you expect to be out after dark, leave a glow stick or light to mark your return target.
12. Items you may want to bring with you include:
    - water and snacks
    - flashlight
    - protective clothing for any possible weather conditions
    - rock carrying backpack or container
    - rock treasure scoop
    - hand lens or other magnifier
    - rock hammer and goggles
    - shovel or other digging tools for excavating
    - water spray bottle
    - UV flashlight (if looking for fluorescent rocks)
    - bug spray or protective clothing
    - cell phone
    - camera (if there is not one on your phone).

Please, appreciate and protect our heritage of natural resources. Conduct yourself in a manner that will add to the stature and public image of rockhounds.

Introduction

After decades of operating the Gitche Gumee Agate and History Museum (located in Grand Marais, Michigan), two things have become clear regarding the objectives of rockhound visitors. Some want to learn to find the elusive Lake Superior agate. Others — especially those impacted by "pretty rock syndrome" — simply want to learn to identify common beach rocks. And, of course, some want to successfully accomplish both objectives. To help rockhounds with agate hunting, over the last 20 years the museum has published three books about agates (two of which are out of print). Until now, the only resource published about beach rocks was a laminated trifold: *Gitche Gumee Beach Rock Identification Guide* (see Figures 1 and 2). This dichotomous key has a decision-making tree wherein the readers answer nontechnical questions and follow the flow chart: green for YES answers; red for NO answers. After answering a few questions, readers are directed to one of the 22 yellow boxes identifying the rock. The yellow boxes correspond to the photographs on the guide's back. Copies of this identification guide (and other books) can be purchased at www.agatelady.com; click on gift shop. Although this identification guide has been helpful, it is clear people want more information.

Figure 1: Rock Identification Guide, front/back of tri-fold.

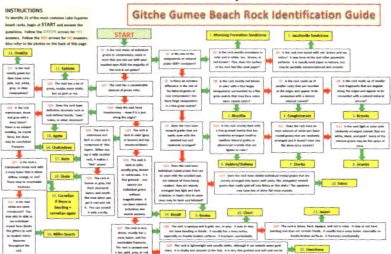

Figure 2: Rock Identification Guide, inside dichotomous key.

During our primary education, all of us learned Earth Science. Unfortunately for most, this information has been long forgotten. Although it is enjoyable to look for "pretty rocks," many want to go to the next level and be able to correctly identify the names of the geologic treasures they find. The process of learning to identify rocks can be modeled as a pyramid. While we may aim to understand one particular topic or concept (the top of the pyramid), this can only be achieved by having a broad base of knowledge that supports the rest of the learning process. This knowledge base gives us a firm and flexible foundation to achieve the educational goal. Hopefully, this book will help rockhounds to build pyramids of understanding about rocks.

Up until now, those who have attempted to learn how identify rocks have had to rely on the rock identification books available on the market. Many buy multiple books to help get up the learning curve. Despite their information-seeking behavior, many find these comprehensive identification books to not be helpful. They feel these books are too jargonistic, contain too few photographs, and include far too many types of rocks and minerals in an attempt to be comprehensive.

This book has been compiled to address these complaints. The main goal is to help people learn about and identify only the most common rocks. In a few cases, minerals that formed in pockets within these rocks are also described. Although some of the rocks, minerals, and geologic history features the Great Lakes region, some of the general information is relevant to specimens found in other geographic regions. There are more than 5,650 different minerals in the Earth's crust. Rocks constitute a combination of minerals. There is no official estimate of the total number of rock varieties since the number of mineral combinations is seemingly endless. Any unique combination of chemical composition, mineral types, crystal grain size, texture, and other characteristics can distinguish and define a rock type. Not that it is an official list, but the Wikipedia web page listing rock types includes nearly 200 varieties (https://en.wikipedia.org/wiki/List_of_rock_types).

The rocks and minerals in this book have been arranged by the methods of geologic formation. By presenting the rocks according to their similar geologic backgrounds, the common formation story for each group can be told (see **Geologic Background** sections in Chapters 2-5). This book features 19 rocks, describes an additional 13 rocks, and mentions 11 minerals that filled in open rock pockets. For each featured rock, information is included about how the rock formed and what characteristics and techniques can be used to identify the rock. Numerous photographs are included for both the featured and described rocks and minerals. Some information is also included for the rocks and minerals that are described (but not featured), but there is less detail. The varieties covered in this book together make up the vast majority of the rocks contained in the Earth's crust and, thus, found in rock hunting areas.

This book can be used in multiple ways to accommodate different levels of information-seeking behavior. To help readers navigate the possibilities, the main chapters have been organized with four color-coded page backgrounds (described on page 3). If you only wish to identify a specimen, begin by skipping ahead in this introduction to the **Rock Identification** section on page 6. Please keep in mind that this book is not intended to be comprehensive. I have included the most common rocks found worldwide. Of the 43 rocks and minerals included, eight are exclusively from the Great Lakes region. These regional rocks serve as examples, especially in cases when there are thousands of varieties.

If you want to learn to understand how rocks formed, begin by reading the **Geologic Background: A Few Basic Facts** beginning on page 4. Continue by reading Chapter 1, which gives a brief description of Earth's geologic history focusing on how the number of minerals on our planet increased from the original number of 60 to over

5,000 minerals! At the beginning of the remaining chapters, there is a **Geologic Background** segment that describes how the rocks in that chapter formed. That segment is followed by a list of featured and described rocks, followed by the rock varieties themselves.

Color Coding Key

NO BACKGROUND COLOR: These sections include basic information relevant to all readers. For each featured or described rock, there is general information about the rock included on pages without a background color. Some of this information does contain geologic information, but these details are important to help in understanding and identifying the rock.

GEOLOGIC BACKGROUND: These sections include geologic information and formation details about the rocks included in that chapter. These sections are highlighted in red.

ROCK IDENTIFICATION: These sections include information about how to identify the featured rocks. Some basic geologic information is included about each rock, but only that which is useful in understanding what you see when you closely examine a specimen. The information is organized into two columns. The left column contains a list of IDENTIFICATION TIPS describing easily observable characteristics and techniques to help readers learn what to look for when examining and classifying a specimen. The right column contains images that illustrate the identification tips. These sections are highlighted in blue.

ROCK PHOTOGRAPHS: These sections include photographs of the rocks to help facilitate classification. Each photograph has a caption describing details about the rock. These sections are highlighted in green.

Notes about the figures included in this book: Each chapter includes numerous photographs and diagrams, all with captions. Over half the figures are my own, but also included are images from NASA, the United States Geologic Survey (USGS), Internet licensing services, and Wikipedia. In cases where there are lots of world-wide variations of a particular rock type, no effort has been made to include photographs of all varieties.

Although I have featured rocks and geologic events from the Lake Superior region, since this is where I live, images are included of varieties found throughout the world. Some of the rocks and minerals featured in this book are only found in the Great Lakes region (see *asterisks in the Table of Contents and the Featured Rocks lists in Chapters 2 to 5). If you live in a different geographic region, you can supplement the information in this book by researching the varieties found in your area. Useful resources include the Internet as well as local rock shops, geology museums, other books, and rock clubs.

All the illustrations and drawings created for this book were generated using simple and easy to understand graphics. When the information was available for magnified images, the size of the specimen or amount of magnification was included. Finally, attribution credits for all outside acquired figures are listed in Appendix F, which begins on page 223 (in order by chapter figure number).

Geologic Background: A Few Basic Facts

To help develop general understanding about rocks, it is important to review a few basic facts. Although this book has been written to be as non-jargonistic as possible, certain terms and concepts must be understood to build the "rock information pyramid." The most basic fact is that minerals make up our planet and everything on it, including us. Minerals are pure substances formed by geological processes. Minerals have definitive chemical compositions, specific crystalline structures, and distinctive properties. If a mineral is divided into fragments, all pieces are the same. For example, quartz is a mineral. When a quartz specimen is broken into smaller fragments, the pieces are still all quartz. The atoms in every piece have the same molecular structure, and all pieces have the same hardness and other properties.

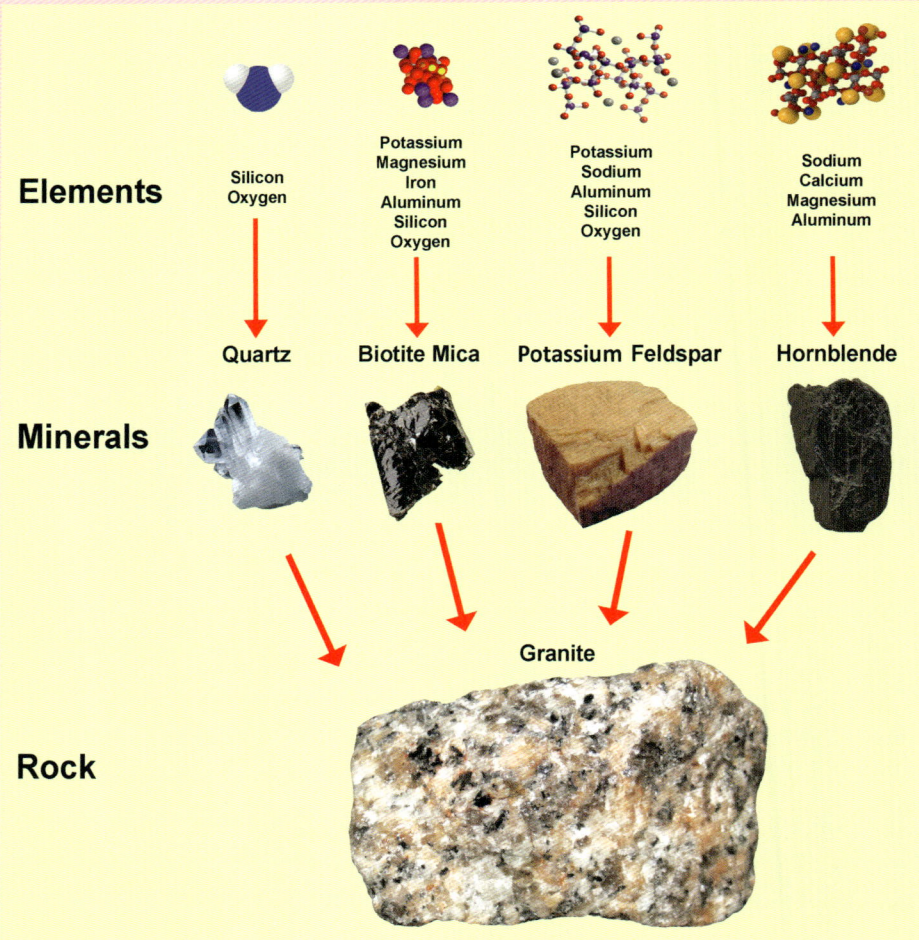

Figure 3: Elements to Minerals to Rocks.

Rocks are also naturally occurring solid substances, but they contain a combination of minerals. Rocks are categorized by how they formed and the minerals they contain. Most rocks are mixtures — they can be separated into individual mineral components with different properties. Thus, if a rock is broken into smaller fragments — the pieces are *not* all the same. Granite, for example, is a rock. It is made of specific elements that formed minerals, which combined into a mixture containing quartz, mica, feldspar, and hornblende (see Figure 3).

Most rock identification books rely on complex, technical characteristics including hardness, color, luster, density, cleavage, specific gravity, refractive index, etc. In this book, color is used to indicate possible variations. Hardness is also used because it is easy to test with common household objects (see Figure 4). Hardness testing is useful to help identify a rock, especially when differentiating one type of rock from another. To test for hardness, as an example, take the specimen and try to scratch a spare piece of glass. After you execute the scratch test, wipe off any residue and examine the glass to see if the rock left a scratch mark. If the rock scratches glass, it is harder than 5.5 on the Mohs Scale (1 is softest; 10 is hardest).

Figure 4: *Mohs scale hardness of common objects.*

Mohs Hardness of Common Objects

Object	Hardness
Fingernail	2.0 - 2.5
Copper	3
Nail	4
Glass	5.5
Knife blade	5.0 - 6.5
Quartz	7

The hardness of knife blades vary, but but if a knife leaves a scratch on a rock, the rock has a hardness of less than 5.0. See Appendix A for more information about how to execute a hardness test. Other than color and hardness, the identification information in this book uses non-technical, observation methods.

Because all minerals are made of crystals, a few terms about crystals must be understood. A mineral crystal is a solid material with elements and atoms arranged into a specific microscopic structure. When the mineral crystals in a rock are large enough to be seen by the human eye without magnification, the rock's texture is "macrocrystalline."

Note: The performance of a hardness test is most useful in identifying minerals. Since rocks are a mixtture of minerals, each of which has its own physical characteristics, the hardness of a rock is only a relative, combined hardness of the rock's mineral components. A pegmatite, for example has quartz (hardness = 7), feldspar (hardness = 5.5), and mica (hardness = 2.5).

Amethyst crystals, for example, are macrocrystalline (see Figure 5). When mineral crystals cannot be seen without magnification, the rock is "microcrystalline." The crystals in agate, for example, are so small they can only be seen with a scanning electron microscope (see Figure 6).

Figure 5: *Macrocrystalline amethyst (crystals can be seen without magnification).*

Figure 6: *Microcrystalline agate (a scanning electron microscope is needed to see these crystals). 200 um=.0078 inch/.2 mm.*

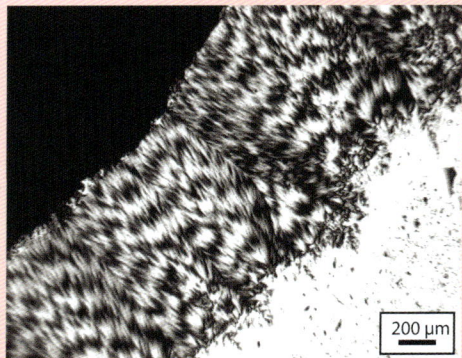

Note: Technically, microcrystalline refers to crystals which are microscopic, hence can be seen with a standard classroom microscope. When the crystals are so small they can only be seen with more powerful magnification (e.g., a scanning electron microscope), the crystal texture is considered to be cryptocrystalline. For communication purposes, the term "microcrystalline" will be used in this book to include ALL crystal textures not visible to the unaided eye.

Note about microscopic images: It is obvious that the microscopic images included in this book will not help rockhounds identify a specimen. These images, though, will help readers to get "inside" a rock in a way they otherwise could not since most people do not have access to sophisticated laboratory microscopes. When the information is available, the scale or degree of magnification is included.

The cross-polarized light images are especially interesting, such as the top right image on the book's cover. Polarized light microscopy produces images at high magnification and resolution that go beyond ordinary optics. Cross-polarized light eliminates glare and allows an unobstructed view of the specimen's thin section. When the thin section slide is placed between two polarizing filters set at right angles to each other, the optical properties of the minerals in the thin section alter the color and intensity of the light as seen by the viewer. Since the optical properties of minerals differ, most rock-forming minerals can be easily identified by using this method. In some cases, other slides are added on top to enhance the color and measurement capability. The mineral thin sections are cut and prepared at 30 microns thick, which is only .001 inches (.03 mm)!

Rock Identification: The Starting Point

An important thing to consider when trying to identify a rock is that any single specimen may actually be different types of rocks fused together. Also, there are many different rocks in each formation category that are similar in appearance. To be absolutely certain about the identification of many specimens, either significant field experience or sophisticated laboratory equipment is required. Because average rockhounds have neither, the rocks and minerals featured in this book include those that are fairly common as well as those that have identifiable characteristics that can be learned by all rockhounds.

To identify a specimen, start by reviewing the characteristics listed below to determine which sub-list of characteristics best describes the rock. Then, proceed to the corresponding chapter. Once you get to the chapter, if you want to skip the geologic background information, proceed to the list of featured rocks and skim through the chapter to see if your rock is included. Use the identification tips on the blue highlighted pages to narrow down your identification classification.

Intrusive Igneous Rocks (page 15):
- contain mineral crystal grains that are visible without magnification (macrocrystalline);
- have a uniform speckled texture throughout with the mineral crystal grains randomly arranged;
- are very hard and solid;
- when testing hardness — will scratch glass;
- are opaque (except for some of the large crystal grains);
- do not contain fossils.

Figure 7: Example of an intrusive igneous rock.

Extrusive Igneous Rocks (page 60):
- are either all or mostly fine-grained. The majority of mineral crystals are not visible without magnification (microcrystalline);
- are very hard and solid;
- when testing hardness — will scratch glass.
- are opaque;
- have a fairly uniform texture, but may have larger embedded crystals;
- may contain empty holes or pockets, or these pockets may have filled in with secondary minerals after formation;
- do not contain fossils.

Figure 8: Example of an extrusive igneous rock.

Amygdaloidal Minerals (page 93)

If the specimen appears to be made from all the same material, rather than from an assemblage of different components, it may be a single mineral rather than a rock. When some rocks formed, they contained hollow spaces that subsequently filled in with minerals. Some of these minerals eroded free from the source rock, remained intact, and can now be found and collected. Amygdaloidal minerals are included in this book because when they are still embedded in their source rocks, they are useful in identifying the source rock. These minerals are also included to help rockhounds to identify them (e.g., agates).

Sedimentary Rocks (page 120):
- are made of individual sediment particles fused together with a natural cement;
- are soft, especially compared to igneous and metamorphic rocks, and can be scratched with a knife;
- contain sediment components that are randomly arranged;
- have sediments that are around the same size (except for conglomerate and breccia);
- may feel like sandpaper when rubbed — the smaller the grain size and the older the rock, the less gritty specimens feel;
- formed in layers from inches to hundreds of feet thick with deposition patterns visible in large rock outcrops, but less observable in hand specimens;
- are opaque;
- may contain fossils;
- are sometimes fluorescent under ultraviolet light (especially fossils).

Figure 9: Example of a sedimentary rock.

Metamorphic Rocks (page 185):
- have obvious bands, streaks, or clumps of different minerals;
- may contain flattened crystal grains;
- can contain layers of granular crystals in between layers of flattened crystals;
- may have large crystals embedded in a fine-grained matrix;
- may be able to be split into slabs;
- are opaque;
- usually do not contain fossils (except, on rare occasions, slate).

Figure 10: Example of a metamorphic rock.

Chapter 1: Source of Minerals

Geologic Background

To tell the story about the rocks and minerals found today in the Earth's crust, we must go back to our planet's beginning. Scientists conclude the Earth formed about 4.56 billion years ago along with the rest of the solar system. In our corner of the Milky Way Galaxy, a large star ended its life as it ran out of fuel and exploded in a catastrophic supernova blast (see Figure 1.1). The star's mass was ejected out into space forming a giant cloud of gas and dust, which began to spin. Our Sun formed in the middle of the dust cloud (circumstellar disk), followed by the planets (see Figure 1.2). Everything in our Solar System is made from star dust, including us!

Figure 1.1: *This supernova remnant W49B is a thousand years old and located 26,000 light years from Earth.*

Figure 1.2: *NASA artist's conception of a circumstellar disk with leftover star dust circling around a newly formed sun.*

Figure 1.3: *The Earth grew into a planet by fusing the cosmic debris in its path around the Sun.*

Four rocky planets developed closest to the Sun and four gas giant planets formed farther out in the Solar System. The Earth is one of the four inner rocky planets that started to take shape when particles of interstellar dust collided, fused, and clumped together into a larger and larger sphere. At this early stage it is thought that the growing

planetesimal (a mini planet) had only around 60 minerals. When the sphere became larger, its expanding mass increased its gravitational pull within the dust cloud (see Figure 1.3). This attracted iron meteorites to the clumping mass, which boosted the number of minerals on our infant planet to around 250. Scientists have estimated it took around 100 million years for our planet to grow from a sphere 6.2 miles in diameter (10 km) to the Earth's current size. During this period of Solar System development, each planet "cleaned" its own orbit by fusing (accreting) all the cosmic debris in its path around the Sun.

Heat from particle collisions during our planet's formation was significant, but not hot enough to melt most of the initial components. As a result, the infant Earth had a relatively uniform composition throughout. Included in this initial mixture, though, was radioactive material left over from the supernova. Over time, radioactive decay increased the temperature of our early world beyond the melting point of its components. Around ten million years after the Earth formed, the blended rocky mixture became molten enough to cause the mineral components to sort into two levels. The denser and heavier iron and nickel minerals migrated inward toward our planet's center to form the core. The lighter minerals formed a deep magma ocean surrounding the core.

Finally, the planet Earth started to take shape. This first geologic eon in our planet's history, the Hadean, lasted from 4.56 to 4.00 billion years ago. Due to the Earth's geologic forces and time, there are no intact rocks or rock outcroppings from the first few hundred million years that survived to the present time. In the absence of Hadean rocks, scientists have relied on indirect evidence to hypothesize what the infant Earth was like. They assumed that for a long time after formation, our planet was geologically violent, bombarded with meteors, and covered with a molten surface. This fiery, Hades-like image gave rise to the period's name.

Until recently there was no direct evidence to challenge this hypothesis. But beginning in the early 1980s, extremely durable mineral crystals, known as zircons, were discovered in Western Australia. Zircons are so durable and resistant to chemical attack that they rarely break down. They survive many geologic events and add additional rings that grow around the original crystal — like tree rings. These minute crystals can be analyzed to determine their age as well as the physical and chemical conditions that existed when they formed. In 2001 more zircons were discovered in Jack Hills, Australia. At 400 micrometers long, these crystals are just a bit larger than the width of five human hairs (see Figure 1.4).

Figure 1.4: This is one of the oldest crystals found on Earth at 4.390 billion years old. Analysis indicates the Earth was cooler and wetter than what scientists had thought. Other crystals from Jack Hills, Australia have been dated to 4.404 billion years old.

JW Valley
University of Wisconsin — Madison

The chemistry of these crystals — specifically the ratio of oxygen isotopes within them (modified atoms with extra mass) — suggests the temperatures on Earth when they formed would have supported liquid water. Their chemistry can also be analyzed to determine age. To date the crystals, scientists measured the radioactive decay of uranium atoms and discovered the oldest zircon crystals from Western Australia to be 4.39 billion years old! Since these zircons eroded free from hardened rock, this evidence suggests that within the first 200 million years of its existence, our planet cooled enough to both make solid crust and have liquid water on its surface. These crystals provide evidence that the surface of the Hadean Earth was cooler and wetter than what scientists had previously believed.

After the ancient zircon crystals were found, scientists have traced back to determine what they think happened on our planet after it formed. They think an early, very thin crust cooled at Earth's surface. Volcanoes developed over much of the planet releasing gases that formed the initial primordial atmosphere. After Earth's surface cooled to a temperature below the boiling point of water, rain began to fall. The water cycle became well-established and it rained constantly for millions and millions of years. Experts think that a shallow sea completely covered our planet just a few hundred million years after formation. The impact of volcanic activity and liquid water caused chemical reactions, which increased the number of minerals in the Earth's crust from 250 to around 500.

But where did our planet's water come from? Scientists think there were multiple sources. Some of the water was trapped within the original rocks that fused to form our infant planet. Water makes up around five percent of the mass in meteorites. Water was also delivered by icy comets and asteroids that subsequently hit the Earth after its formation. The inner planets of our Solar System were pummeled by these icy objects from 4.5 billion to 3.8 billion years ago (see Figure 1.5). Evidence for these events, known as the Late Heavy Bombardment, is carved into the surface of the Moon (see Figure 1.6).

Figure 1.5: *Water was added to our planet from the comets and asteroids that hit the early Earth.*

Figure 1.6: *This photo of the Moon taken by Apollo 15 shows craters formed from meteor and comet impacts.*

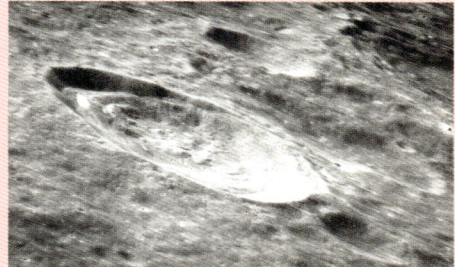

New research suggests there was a third source that delivered water to our planet. Over the last several decades scientists have not only been able to measure the chemistry of the Earth's crust, but they have also investigated the chemistry of our planet's interior mantle that lies below the crust. This analysis measured certain isotopes (modified atoms) to determine whether the material making up the mantle did in fact form in the inner Solar System, as was hypothesized. We know the inner Solar System's planets are more rocky and drier compared to the icy outer Solar System's planets. These distinct halves of our Solar System created markedly different isotopes. When researchers examined isotopes in the Earth's mantle, they discovered that our planet has both types. This was a complete surprise since the Earth should only have isotopes corresponding to the inner Solar System's variety.

To explain this surprising find, scientists have hypothesized that material from the outer Solar System somehow was delivered to Earth after our planet formed. A possible explanation is that around 600,000 years after formation, a Mars-sized protoplanet hit Earth. It is thought this protoplanet, named Theia, originated in the outer Solar System and therefore contained that variety of isotope. Theia was propelled by gravitational forces into the inner Solar System where it collided with Earth (see Figure 1.7).

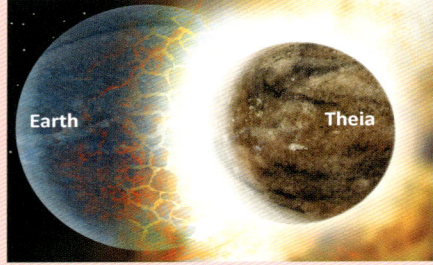

Figure 1.7: *Artist's drawing of the hypothesized collision between Earth and Theia around four billion years ago.*

It is believed that Theia did not hit the Earth head-on, but instead hit with a glancing blow. There were multiple consequences of this impact angle. First, when the impact occurred, only material from the mantle was ejected into space, leaving the Earth's core of iron and nickel intact. Some of Theia's mass became incorporated into the Earth's mantle, while the remaining portion combined with the material ejected from the Earth's mantle to form the Moon (see Figure 1.8). The impact is also thought to have changed Earth's axis to produce the significant 23.5° axial tilt that is now responsible for Earth's seasons. The impact may have also sped up Earth's rotation.

Figure 1.8: *The Moon formed when the planet, Theia, collided with Earth.*

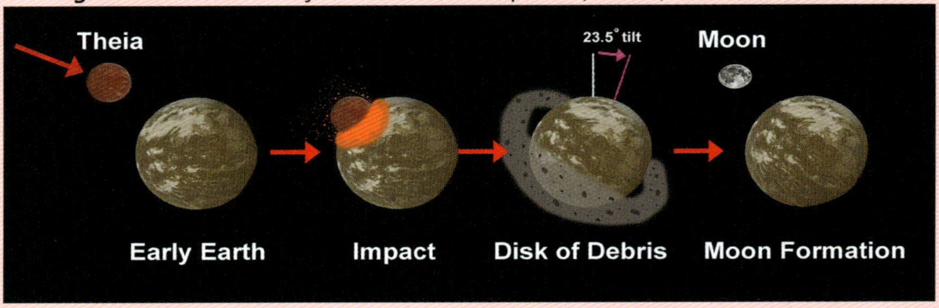

The part of Theia incorporated into our planet appears to have delivered water. Isotype studies verify that some of Earth's mantle came from the outer Solar System where ice (water) is more prevalent. According to the calculations made by scientists, the Theia collision could have provided sufficient material to account for the entire amount of water existing on Earth today. By the end of the period of bombardment and the end of the Hadean Eon, oceans on Earth had formed.

In addition to the formation of oceans during the Hadean Eon, volcanic activity worldwide continued to release tremendous amounts of lava and gases. More and more of the molten lava cooled and hardened, building up with subsequent eruptions (see Figure 1.9). Thus, igneous rock was the first rock on Earth — all rocks existing today have come from these pioneer rocks. Because of the dramatic geologic activities during the Hadean Eon, the number of minerals on Earth increased from 500 to around 2,500.

Figure 1.9: Solidified lava from the Earth's early volcanoes created the first rock on our planet's surface.

As the number of lava flows on the early Earth increased, chains of volcanic islands formed all over the planet. There was a continuous battle between volcanoes trying to create larger and larger islands, and the forces trying to tear them down. The erosion of these islands at first was dramatic and swift for many reasons. The rock had to withstand plummeting rain. Life had not yet evolved so there was no vegetation to protect the rock. Waves from the surrounding shallow seas continuously pounded the rocky islands from all sides. Another significant impact on the rate of rock erosion is a bit more surprising to us living today. When the Moon first formed from the remnants of the Theia collision, it was very close to Earth – perhaps only 15,000 to 20,000 miles away (24,140 to 32,187 km)! Since we get significant tides today caused by the Moon that is now 240,000 miles away (386,243 km), imagine what the tidal action would have imparted on the volcanic islands when the Moon was much closer! The Moon today continues to move away from Earth at a rate of 1.48 inches per year (3.78 cm).

Over time, landmasses began to withstand erosional forces and continued growing to create continents. Perhaps the biggest impact on both our planet and the number of minerals we have in the Earth's crust was biologic evolution. The first organisms on Earth were probably cyanobacteria (see Figure 1.10), which are thought to have evolved over 3.5 billion years ago in shallow seas surrounding the early continents. These microbes survived using photosynthesis. They used

Figure 1.10: Cyanobacteria were responsible for producing free oxygen on early Earth.

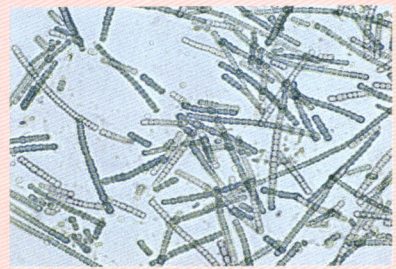

sunshine, water, and carbon dioxide to produce carbohydrates for food, giving off oxygen gas as a byproduct. Prior to the evolution of cyanobacteria, there was almost no free oxygen on Earth or in our atmosphere. Because of the lack of oxygen, the early atmosphere was orange (due to high concentrations of methane and other gases), and the oceans were green (due to the high concentration of uncombined iron molecules). Free oxygen is not stable and will chemically react to form compounds with nearly every element in the periodic table. Once free oxygen became prevalent in the Earth's atmosphere around 2.33 billion years ago, everything changed. By the time the amount of oxygen on Earth reached modern levels, the number of minerals on our planet increased from 2,500 to over 4,300, causing both the ocean and atmosphere to turn blue.

As of December 2020, the International Mineralogical Society (IMA) recognized over 5,650 minerals in the Earth's crust. Around 200 of the listed minerals are common; only 20 are very common. This list includes 1,159 minerals discovered prior to the founding of the IMA in 1959 (grandfathered in), and 96 questionable minerals that are still under investigation. Presently, each year between 90 and 110 new mineral varieties are discovered and added to the list. Even though the number of minerals continue to increase, just ten minerals make up more than 95 percent of the crust (see Figure 1.11).

This book explains how minerals combined, crystallized, broke apart, melted, and recrystallized to form the igneous, sedimentary, and metamorphic rocks found today in the Earth's crust. The more we understand the components and formation processes of rocks, the more success and enjoyment we will have finding and identifying Earth's mineral treasures. Also, the more we understand and appreciate our planet's geologic wonders, perhaps the better we will take care of our celestial home.

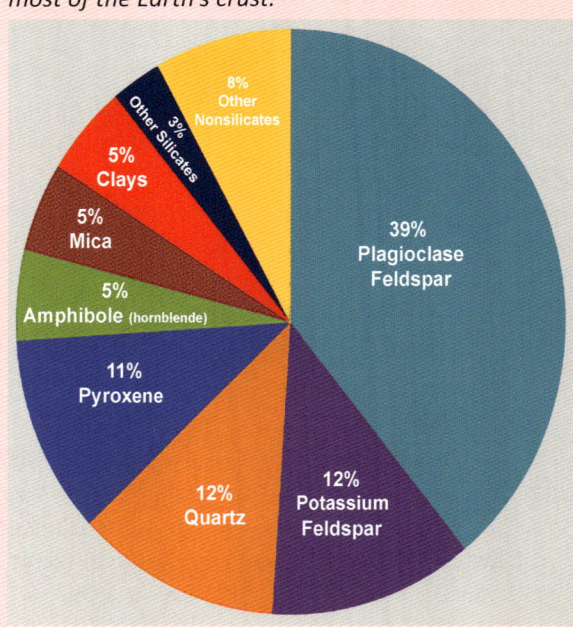

Figure 1.11: A limited number of minerals make up most of the Earth's crust.

Chapter 2: Intrusive Igneous Rocks

Note: For a list of the Featured Rocks included in this chapter, please go to page 21.

Geologic Background

As oceans formed and spread out over the early Earth's surface, water leaked through cracks in the bottom of the thin ocean crust and became superheated by magma. The semi-molten magma and hydrothermal fluids worked together to melt basaltic rock lining the bottom of the ocean's crust. This added new minerals to the melt, which later cooled and recrystallized to form the first granite rock on Earth. The newly formed rock was lighter than basalt, so it "floated" on top expanding the landmasses all over our planet. It was not long before mini continents, called cratons, began to form. Volcanoes continued to spew lava and ash, causing the cratons to enlarge further and spread out. In the center of the cratons, granitic magma intrusions forced their way to the surface, increasing the height and thickness of the craton landmasses. Scientists estimate the average rate of

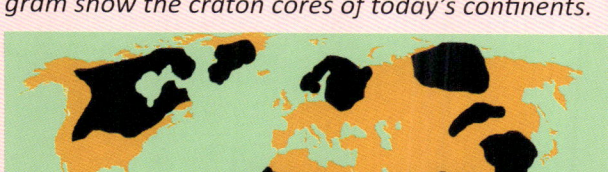

Figure 2.1: *The black highlighted areas in this diagram show the craton cores of today's continents.*

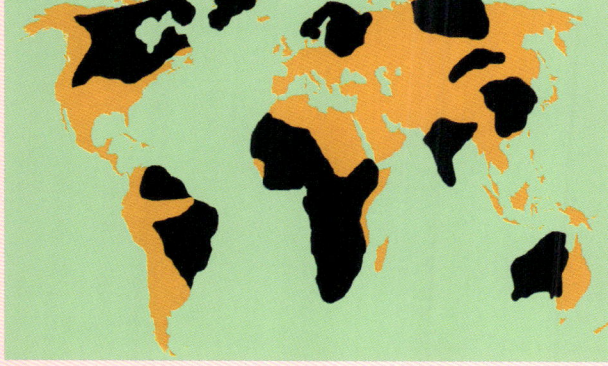

landmass growth for cratons during this period was around one cubic mile a year (4.2 cubic km). The process of rock building was well underway by the end of the Hadean Eon 4.0 billion years ago. These cratons formed the cores around which today's continents grew (see Figure 2.1).

All igneous rocks formed from either magma (molten rock under the Earth's surface) or lava (molten rock above the Earth's surface). This molten rock was (and still is) a complex liquid that varied greatly in composition and properties. Magma has temperatures as great as 2,552°F (1,400°C) and often originated in regions 31 to 124 miles below the Earth's surface (50 to 200 km). Magma ranged from being entirely liquid to partially crystalline (rocky). Because semi-molten magma had a lower density than the solid rock surrounding it in the upper mantle and crust, its higher buoyancy caused magma to seek cracks and other ways to move up toward our planet's surface. The race between upward movement and cooling determined which type of igneous rock formed. When magma cooled and crystallized below the Earth's surface, it formed intrusive igneous rocks (also called plutonic). When magma reached the planet's surface as lava, it solidified as extrusive igneous rocks (also called volcanic). Intrusive rocks are covered in this chapter. Extrusive igneous rocks are covered in Chapter 3.

As magma pushed up from the Earth's mantle through cracks toward the planet's surface, it sometimes became trapped by impenetrable rock thousands of feet below the surface. Eventually the magma cooled and its progress toward the Earth's surface was stopped, where it crystallized into intrusive igneous rock. These deep "pools" of magma are called intrusions. Three common types of intrusions are sills, dikes, and batholiths (see Figure 2.2). All the rocks featured in this chapter formed in an intruded body of magma.

Figure 2.2: *Examples of magma intrusions.*

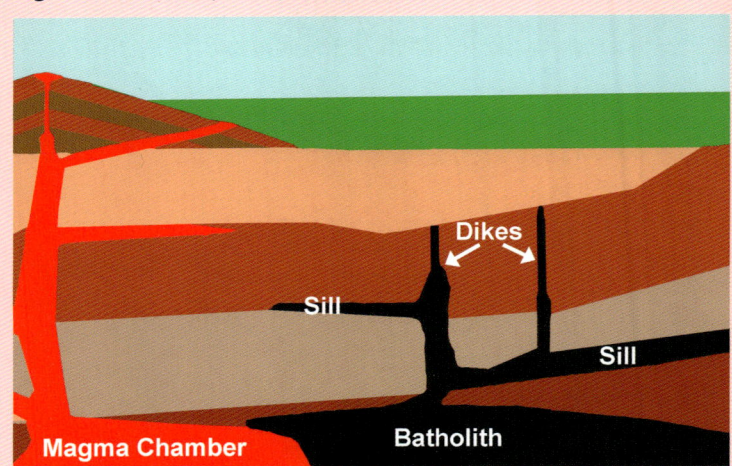

Sills formed when magma intruded between different layers of preexisting rock, forming a horizontal or gently sloped shelf of igneous rock. Intrusive sill formations did not cut across rock layers but squeezed in between them (see Figure 2.3).

Dikes formed when magma pushed up towards the surface through vertical cracks in existing rock. They either cut through multiple layers of existing rock or through a large continuous layer. Dikes are narrow and long in shape. They are always younger in age than the rock through which they intruded (see Figure 2.4).

Batholiths are large, deep-seated intrusions (sometimes called plutons) that formed when large pools of magma slowly made their way toward the surface but became trapped 3 to 18 miles (5 to 30 km) below the surface. To be classified as a batholith, the intrusive rock formation must be at least 40 square miles in size (100 km^2). The batholith in Figure 2.5, located in Yosemite Valley, was uplifted to the Earth's surface by geologic forces.

Figure 2.3: Intrusive sill formation on Mount Gould, Glacier National Park, Montana.

Figure 2.4: The basalt dikes cut through host rock located along the Baranof Cross-Island Trail, Alaska.

Figure 2.5: Large batholith mountains were uplifted by plate tectonic forces in Yosemite Valley, California.

So just how did crystals form in magma to make igneous rock? When this semi-molten fluid was hot, the atoms were always in motion. Magma stayed molten because the heat provided energy to the atoms (kinetic energy) which ensured they would keep moving and not begin to crystallize. Some atoms may have collided and created temporary bonds before breaking apart, but the magma stayed molten. The molten stage stayed steady at first because it was hot enough to have a balance between the formation of atomic bonds and the rate at which they broke apart. As magma cooled there was less kinetic energy, so the atoms slowed down. Eventually, when magma sufficiently cooled, atoms slowed down enough so that some molecular bonds between atoms persisted. The first molecular bonds served as small nuclei, which acted as the centers around which additional molecules attached (see Figure 2.6). Over time, solidified nuclei crystals continued growing until they merged to become full macrocrystalline crystals.

Figure 2.6: Diagram of the crystal formation process.

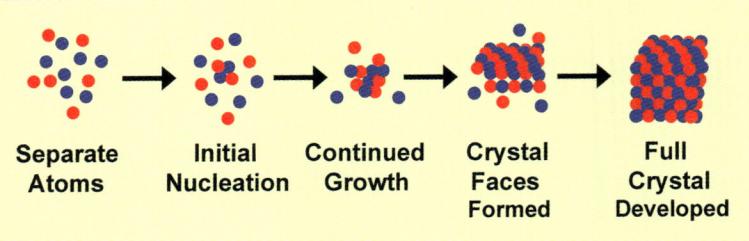

Because intrusive igneous rocks we find today on the Earth's surface solidified from a semi-fluid, well-mixed magma, they tend to have uniform structure with the mineral grains evenly and randomly arranged (see Figure 2.7). The texture of intrusive igneous rocks is a crystalline structure with an interlocking mosaic of mineral crystals that fit together like pieces of a jigsaw puzzle (see Figure 2.8). There was no "gluing" agent required to hold the mineral grains together since the crystals instead chemically bonded on the molecular level.

Figure 2.7: *Example of intrusive igneous rock with uniform, randomly arranged mineral grain crystals.*

Figure 2.8: *This cross-polarized microscopic image of an intrusive igneous rock shows chemically bonded mineral grains.*

There were several factors that controlled crystal growth in igneous rocks. The most important factors were temperature, time, abundance of necessary elements, and the presence or absence of a flux. High (but not too high) temperatures fueled atoms with kinetic energy that allowed them to migrate through the magma and attach to the active crystallizing surfaces surrounding the pioneering nuclei. Time allowed more atoms to migrate to the growing crystal surfaces, increasing the size of the crystal grains. And finally, in cases where there was an influx of either hot hydrothermal fluids or additional magma, these liquids served as a flux to deliver more kinetic energy and more atoms to the crystallization sites, thus, growing the mineral crystal grains even larger.

Intrusive igneous rocks are made up of different mineral crystals that grew together and combined within the semi-molten melt as it cooled. Under the Earth's surface, crystals grew slowly in the magma because of the insulating influence of the surrounding rock, which slowed down the magma's cooling process. When magma cooled slowly, the resulting crystals in the rock grew large enough in size to be seen without magnification (see Figure 2.9). Crystal grains are considered coarse-grained when they are more than .20 inches in size (5 mm). Coarse-grained intrusive rocks tended to form in large batholith intrusions. In some intrusive rocks, crystals have been found that are over ten feet in length, weighing over a ton (3 m, 907 kg). Fine- and medium-grained intrusive rocks with crystals smaller than .20 inches in size (5 mm) tended to form in smaller intrusions such as sills and dikes.

Figure 2.9: Granite has a coarse-grained texture with mineral crystal grains that are visible with the unaided eye.

Some previously deeply buried intrusive igneous rocks have now been exposed on our planet's surface. This happened in at least two ways. Either there were long periods of weathering and erosion that removed overbearing rock and soil, or plate tectonic forces uplifted the intrusive rock to the surface, or both. Over geologic time, many batholiths have been uplifted to the Earth's surface and exposed as mountains. Yosemite's Half Dome and Mount Rushmore are both parts of uplifted batholith intrusions (see Figures 2.10 and 2.11).

Figure 2.10: California's Yosemite's Half Dome batholith intrusion has been uplifted to the Earth's surface by geologic forces.

Figure 2.11: The granite comprising South Dakota's Mount Rushmore is made from an uplifted batholith.

Featured Rocks

Intrusive igneous rocks are classified based on the proportion of their mineral components. It is difficult to determine these percentages without significant field experience, and to be certain, without measuring the component percentages using sophisticated laboratory equipment. Because most rockhounds do not have either of these advantages, it can be difficult, frustrating, and almost impossible to identify rocks with absolute certainty. There are more types of intrusive igneous rocks than will be covered here. Those featured in this chapter include the most common, as well as those that can be identified using observable characteristics. Rocks covered in this chapter include the following:

> **Key**
> - Featured rock (includes rock identification information and photographs).
> - Rock is described but not featured (includes photographs).

- **Granite** (page 22): the most common intrusive igneous rock;

- **Pegmatite** (page 32): an intrusive igneous rock with large crystals;

- **Granodiorite** (page 38): an intrusive intermediate between granite and diorite. This rock will be described but not featured;

- **Diorite** (page 39): an intrusive intermediate between granite and gabbro;

- **Gabbro** (page 46): a coarse-grained rock that is the intrusive equivalent to extrusive basalt. **Diabase** is so similar to gabbro — it has slightly smaller crystals — that these two intrusive rocks have been grouped together;

- **Syenite** (page 54): a granite-like rock that contains less quartz. Only the fluorescent variety will be described.

Granite

Granite is the most common intrusive igneous rock in the Earth's crust. Continents are the exclusive home of granitic rocks. Many mountain ranges such as the Andes, Himalayas, Rockies, and Sierra Nevadas are composed of giant granite intrusions that have been uplifted to the Earth's surface by plate tectonic forces (see Figure 2.12).

Figure 2.12: Mount Whitney, part of the Sierra Nevada mountain range in California, is comprised mostly of granite. It is the tallest mountain in the continental United States at 14,505 feet (4,421 m).

According to the theory of plate tectonics, granitic continents are the only part of Earth's crust not recycled during the plate tectonic cycle. Because granite is a relatively lightweight rock, continents stay intact floating on the Earth's surface to form anchors for large tectonic plates (see Figure 2.13). These plates are made up of both continental crust and oceanic crust. There are seven large tectonic plates and six minor plates. Granite mountains are uplifted by the force of two tectonic plates colliding together. When this happens, one tectonic plate loses the collision battle and subducts (sinks) under the other plate. These collisions happen because tectonic plates do not stay in one place. Scientists determined in the mid-twentieth century that continental plates move around the planet one to two inches per year (3 to 5 cm), driven by hot convection currents churning in our planet's mantle (see Figure 2.14). Because continents "float" at the Earth's surface and are not subducted and destroyed by plate tectonic processes, they provide a form of memory because the records of geologic processes, climate change, and biologic evolution can be read in their rocks.

Scientists think tectonic processes began on our planet between 3.3 and 3.5 billion years ago. Although some granite has been uplifted to the surface, most is still buried. Nearly everywhere on any continent, no matter what rocks are at the surface, it is possible to drill down and reach granitic rocks. Granite intrusions may be several miles thick and just as wide.

Figure 2.13: Major and minor tectonic plates.

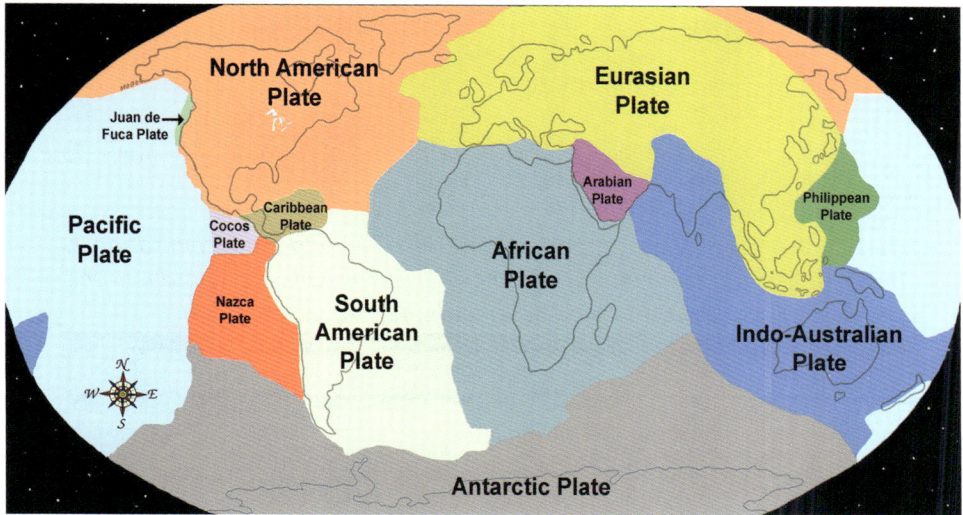

Figure 2.14: Convection currents in the Earth's mantle drive plate tectonic forces.

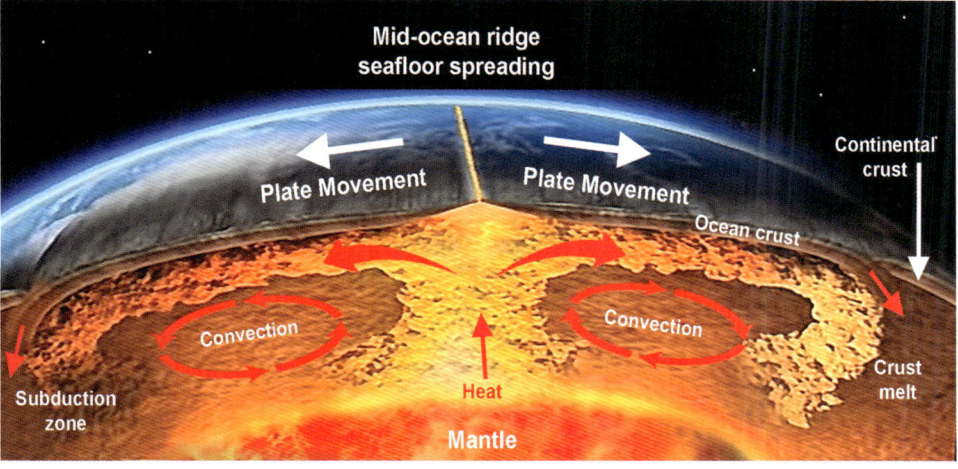

There are numerous varieties of intrusive igneous rocks that look very much alike. Because of the difficulty in identifying the varieties of rock in the "granite family," these related specimens are collectively called granitoids or granitic rock. The difference between granitoid rocks is the proportion variations of mineral components. Another complicating factor involving the identification of granite is that the word "granite" has different meanings, depending on who is using the term. According to geologists, granite is limited to intrusive igneous rocks consisting of at least 80 percent quartz and feldspar combined, with quartz comprising 20 to 50 percent of the specimen. However, commercial companies selling countertops, tile, and other construction materials label any rock with visible crystal grains as "granite." This broader definition includes other types of igneous as well as metamorphic rocks.

In geology, granite is a light-colored, medium- to coarse-grained intrusive igneous rock with easily seen crystals that are randomly and evenly arranged throughout the specimen. Granite consists mostly of quartz and feldspar with scattered darker biotite mica, hornblende crystals, and sometimes shiny silvery muscovite mica crystals peppered throughout (see Figures 2.15 to 2.19). These minerals are also components in many of the other rocks included in this book. The inter-grown mineral crystals are all about the same size — a characteristic of slow cooling under the Earth's surface. Because the exact mineral percentages change from one granite specimen to the next, the color of granite varies. Most of the time, granite is light in color with a mix of white, pink, gray, and black.

Figure 2.15 to 2:19: *Quartz, feldspar, biotite mica, hornblende, and muscovite mica.*

Granite rock can be found worldwide. All granite specimens originated from mountains that were uplifted and exposed to weathering and erosional forces. Mechanical weathering split large granite outcrops into smaller and smaller fragments. The granite boulder shown in Figure 2.20 was split by frost. Additional weathering forces separated the granite fragments into its different mineral components. Chemical weathering broke down the softer components, including the feldspars and micas, into clay minerals (see Figure 2.21). Soluble components dissolved and were carried away by groundwater or surface waters. The primary mineral in granite — quartz — resisted weathering, remained virtually unchanged, and became gravel, sand, and silt. A magnified image of quartz-dominated beach sand is shown in Figure 2.22.

Figure 2.20: *Granite boulder mechanically weathered and split by frost.*

Figure 2.21: Clay minerals formed from the chemical weathering of feldspar and mica. This scanning electron microscope image is magnified 1,340 times.

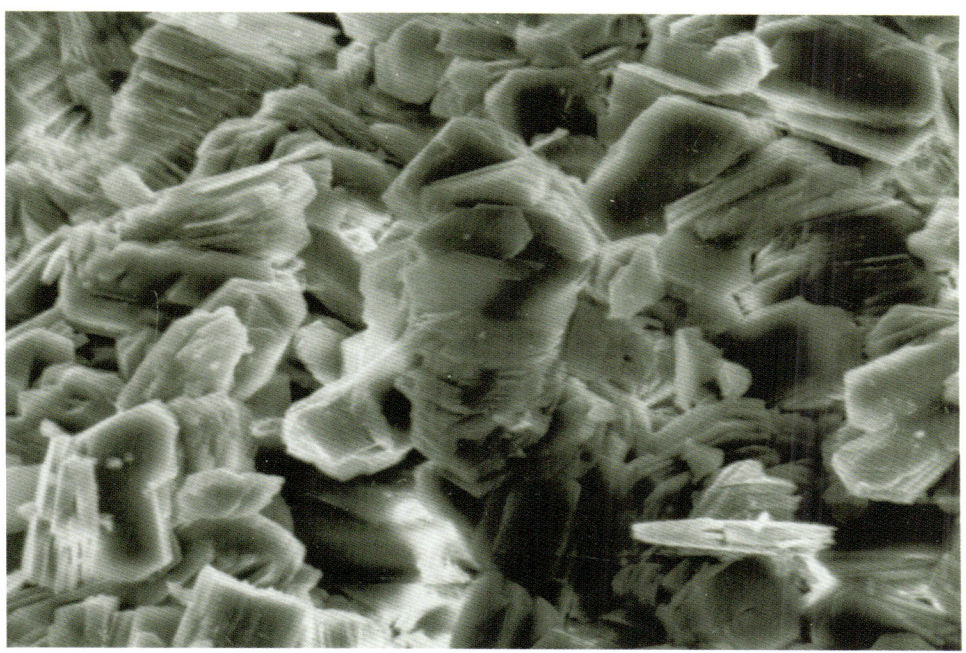

Figure 2.22: Durable quartz beach sand resulted from the weathering of granite. The sand in this image is magnified around 300 times.

Rock Identification: Granite

Because there are so many similar coarse-grained intrusive igneous rocks, it may be best (especially for beginners) to identify specimens as being in the "granitoid" family of rocks. To determine if a specimen meets the geologist's definition of granite, the following statements must be true.

Rock Identification Tips	Photographs or other identification information
Mineral crystal grains can be seen without magnification.	**Figure 2.23:** The crystal grains in granite are visible to the unaided eye. 
Mineral crystal grains are arranged in a speckled pattern that is mostly random and uniform throughout the specimen.	**Figure 2.24:** Granite has a random and uniform arrangement of crystal grains. 
Feldspar minerals give granite most of its color, which may be speckled with white, gray, black, and pink mineral grains. Granite is mostly light in color.	**Figure 2.25:** Granite is light in color but has a mixture of light and dark minerals. Because the exact proportions of minerals change from one granite specimen to the next, granite can range widely in color, crystal grain size, and other characteristics. 

Granite has a hardness greater than 5.5 on the Mohs Scale and will scratch glass.	*Figure 2.26: Granite is a hard rock that will scratch glass.*
Granite is opaque.	*Use a bright flashlight to verify that light does not shine through the specimen.*
Granite never contains fossils.	*Since granite formed below the Earth's surface from semi-molten magma, it does not contain fossils.*
Use magnification to verify there are feldspar and quartz crystal grains. Also, look for mica and hornblende crystals.	*See Figure 2.27 below. This picture shows Columbia Pink Granite from Maine.*

Figure 2.27: The mineral grains in granite are shown in this photograph and described on the next page.

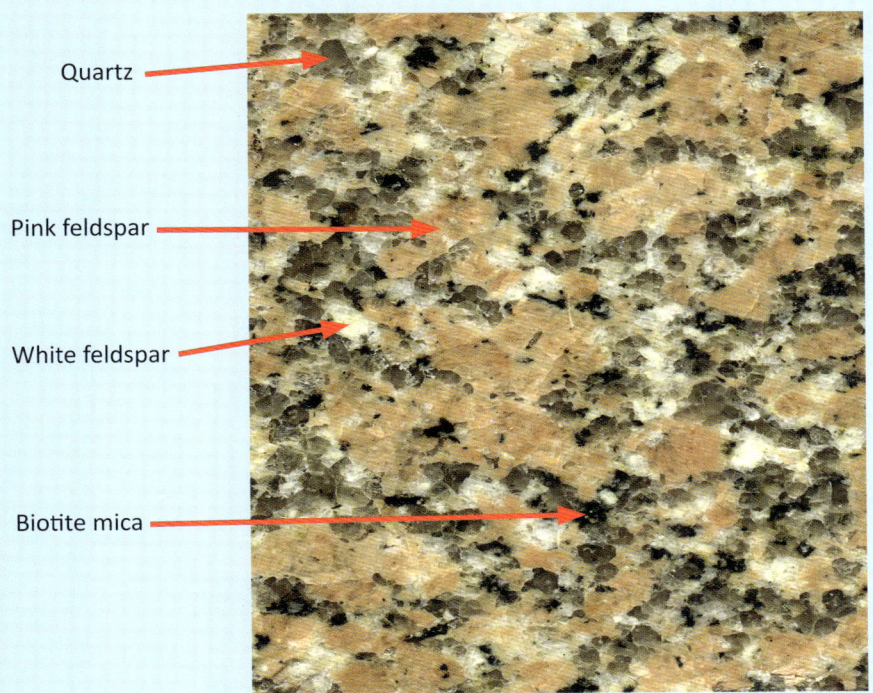

27

Feldspar When the magma cooled enough for crystals to start forming, feldspar crystals took their turn first. Feldspar crystallized at high temperatures and because these crystals formed first, there were no obstructions. They had plenty of time and space to grow, allowing them to develop intact crystals with straight sides. Feldspar crystals are often blocky in shape and can have linear striations etched across the flat-sided crystal grains that resemble scratches or fractures. When you hold and rotate a granite specimen in full sun or in a bright light, the flat crystal faces will glint up at you. Potassium feldspar can be pink or white and has a porcelain luster. Less common is plagioclase feldspar, which is gray or white in granite. Striations along the crystal faces are more common in plagioclase feldspar.

Quartz These crystal grains formed after the feldspar crystals and filled spaces between the already formed crystals. As a result, quartz crystal grains appear as irregular blobs crowded up against the straight-sided feldspar crystals. Quartz grains in granite appear glassy and can be gray, white, or almost clear.

Mica Two types of mica crystals are often in granite. *Muscovite* crystals are flat flakes that can be clear, brassy, or gray. These crystal flakes developed in flat layers. The flat crystals reflect light when the specimen is rotated in sunlight or in a bright light. Some people mistake muscovite mica crystals for gold or silver. With magnification, use a sharp knife to verify that the layers of mica can be scratched. Mica layers can also be split apart and flaked off. *Biotite* mica crystal flakes also formed in layers. These mica flakes, though, are black.

Hornblende These dark crystals are obvious in granite, but not all granite samples have them (such as this sample). When hornblende crystallized in magma, it formed long, black crystal grains that appear as distinct, rod-shaped crystals — sometimes with pointed ends.

Granite Photographs

Figure 2.28: Most granites are quite light in color.

Figure 2.29: This granite has large crystal grains, so it grades into pegmatite (see next rock in this chapter).

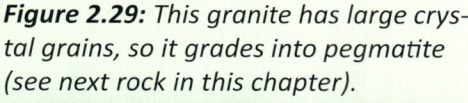

Figure 2.30: This granite specimen has a lot of quartz.

Figure 2.31: A colorful granite.

Figure 2.32: This granite specimen cooled slowly, allowing large crystals to form.

Figure 2.33: This Colorado granite is 1.42 billion years old.

Figure 2.34: A granite specimen on a Lake Superior beach.

Figure 2.35: Hornblende at times crystallized first, resulting in a linear arrangement of dark crystals with a streaky, gneiss-like appearance.

Figure 2.36: Some granites have almost no pink potassium feldspar. These specimens are difficult to discern from other granitic rocks.

Figure 2.37: This rough granite specimen shows random crystal arrangement.

Figure 2.38: This Quebec granite specimen has larger than normal feldspar crystals.

Figure 2.39: A close-up of a granite specimen from Scotland.

Figure 2.40: This Minnesota granite formed 1.75 billion years ago. Top to bottom view = 9.3 inches (23.7 cm).

Figure 2.41: Another close-up image of a typical granite.

Figure 2.42: This granite specimen had fractures that subsequently filled with extrusive igneous basalt (secondary fill).

Figure 2.43: This rock is a mixture of granite (top) and gabbro (bottom).

Pegmatite

Pegmatites are a variety of granite with the largest crystals in the granitic family of rocks, usually larger than a half inch (1.25 cm). Extremely large crystals have been found in pegmatites, including some over 30 feet in diameter (10 m). Pegmatites formed below the Earth's surface at the top or along the outer edges of magma chambers, or in smaller intrusions leading out of the magma chambers. This rock developed during the final stage of crystallization within magma chambers, so it was the last rock to solidify. Like granite, most minerals in pegmatites include quartz, feldspar, and mica. Rare minerals are also found in pegmatites such as beryl (emerald and aquamarine), tourmaline, topaz, fluorite, apatite, corundum (ruby), tin, tungsten, and others. The molecules for these rare minerals had such low concentrations in the original magma melt that there were not enough molecules to crystallize and become incorporated in other intrusive igneous rocks. At the end of rock solidification in the magma chambers, molecules for these rare minerals became concentrated to the level that allowed crystallization (see Figure 2.44).

Figure 2.44: *This pegmatite boulder has large emerald crystals*

In most intrusive igneous rocks, larger crystals formed when significant time was available to allow mineral molecules to migrate and chemically bond onto active crystal growth surfaces. With pegmatites, large crystals formed instead because of the low-viscosity fluids (thin, flowed easily) that collected at the top of magma chambers. During most of the time crystals formed inside the igneous intrusion, water was not removed from the magma. Like the rare mineral molecules that accumulated, the water concentration increased until there was an overabundance of water at the top of the magma chamber. Over time, the water separated from the remaining magma. Pockets of superheated water accumulated that were extremely rich in dissolved minerals. Since the water was much less viscous than the magma, the mineral atoms were more mobile allowing them to freely move through the water and rapidly bond to form and grow large crystals in the pegmatite rock (see Figure 2.45).

Figure 2.45: *This Crabtree Pegmatite from North Carolina has black schorl tourmaline crystals, pinkish-reddish garnet crystals, medium-gray quartz, and whitish-gray feldspar. This deposit from the Devonian period is around 400 million years old.*

Rock Identification: Pegmatite

Pegmatite rocks are not as common as granite and other intrusive igneous rocks. A pegmatite specimen can be identified by the following characteristics.

Rock Identification Tips	Photographs or other identification information
Pegmatites consist of large inter-locking crystals that are chemically bonded at the molecular level. Most of the crystals are at least a half inch in size (1.25 cm) — some crystals may be larger.	**Figure 2.46:** This cross-polarized microscopic image of a Norway pegmatite shows large mica and feldspar crystals. Field of view = .13 inches wide (3.3 mm). 
Most of the mineral crystals in pegmatites are quartz, feldspar, and mica, but there may also be other crystals made of minerals not found in granite. Some of these other crystals may be precious gems and minerals.	**Figure 2.47:** This pegmatite has coarse pink potassium feldspar, clear gray quartz, and sparkly muscovite and biotite mica. 
The mineral crystals in pegmatites may not be uniform in size as they are in granite. Pegmatites usually have different zones of crystallization with a variety of mineral sizes and assemblages.	**Figure 2.48:** This pegmatite specimen has different sized crystals. 

The different mineral crystals in pegmatites may not be equally distributed throughout the specimen.	*Figure 2.49:* This pegmatite shows uneven distribution of crystals. 
Pegmatite specimens have a hardness of 6.0 on the Mohs Scale and will scratch glass.	*Figure 2.50:* Pegmatite rocks will scratch glass. 
Pegmatites are opaque (other than some of the large crystals).	Use a bright flashlight to verify that light does not shine through the specimen.
Pegmatites never contain fossils.	Since pegmatites formed below the Earth's surface from semi-molten magma and hydrothermal solutions, they do not contain fossils.

Pegmatite Photographs

Figure 2.51: An Australian pegmatite.

Figure 2.52: A Colorado pegmatite with large feldspar and quartz crystal (1.4 billion years old).

Figure 2.53: A Brazilian pegmatite with feldspar and mica.

Figure 2.54: This pegmatite boulder from Wisconsin has course pegmatite bands in granite.

Figure 2.55: Pegmatite specimen from a Lake Superior beach.

Figure 2.56: South Dakota pegmatite with white potassium feldspar, gray quartz, and muscovite mica.

Figure 2.57: A pegmatite from the Gitche Gumee Museum's collection.

Figure 2.58: This rock is intermediate between granite and pegmatite.

Figure 2.59: This Lake Superior pegmatite is another specimen intermediate between granite and pegmatite.

Figure 2.60: Close-up of a pegmatite specimen.

Granodiorite

Granodiorite is an intrusive igneous rock similar to granite that formed from silica-rich magma, but the percentage of primary feldspar is opposite than what is in granite. Granodiorite contains more plagioclase feldspar, whereas granite contains more potassium feldspar. Granodiorite contains more than 20 percent quartz by volume, and between 65 and 90 percent of the feldspar must be plagioclase. Black hornblende crystals are more prevalent in granodiorite than in granite, giving it a distinctive two-toned and overall darker appearance. This rock can vary from being medium to coarse grained. It is intermediate in composition between diorite and granite. Like granite, granodiorite is the product of the melting of continental rocks near subduction zones where one tectonic plate slipped below another. Huge batholithic intrusions developed, some of which crystallized into granodiorite. Outcrops of granodiorite at the Earth's surface are now exposed after uplift (caused by plate tectonic forces) and erosion occurred. Sometimes granodiorite and other igneous rocks developed orbicular spherical masses called "orbs" or "orbicles." In this case, the rock is an orbicular granodiorite. The orbs can be composed of plagioclase feldspar, hornblende, biotite mica, and magnetite (see Figures 2.61 and 2.62). To have certainty in differentiating granodiorite from granite, laboratory equipment is required.

Granodiorite Photographs

Figure 2.61: An orbicular granodiorite from Australia.

Figure 2.62: Another orbicular granodiorite.

Figure 2.63: Close-up of granodiorite from Sequoia National Park in California.

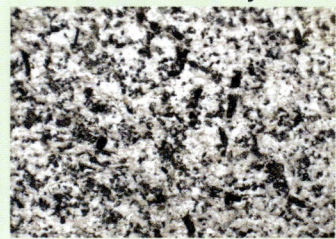

Figure 2.64: Another granodiorite.

Figure 2.65: A cross-polarized image of granodiorite (magnified X 40).

Diorite

Diorite is a medium- to coarse-grained intrusive igneous rock composed of plagioclase feldspar with lesser amounts of hornblende and biotite mica. Diorite is an intermediate between granite and gabbro. It usually contains less than five percent quartz, less than ten percent potassium feldspar, and can have minute amounts of muscovite mica. The crystal grains in diorite are usually all about the same size, which can easily be seen with the unaided eye. Diorite has a salt-and-pepper appearance with a randomly arranged mixture of light and dark crystal grains. However, some diorite specimens show a semi-streaky arrangement of feldspar, biotite mica, or hornblende crystals that have a somewhat parallel arrangement (see Figure 2.66).

Figure 2.66: This diorite specimen shows a somewhat parallel arrangement of the crystal grains.

Diorite formed above subduction zones at convergent plate boundaries where ocean plates subducted beneath continental plates at a rate of one to three inches per year (2 to 8 cm), (see Figure 2.67). Once the ocean plate sank deep enough, it released water into the mantle rock. Water caused the rock to melt at a lower temperature because the bonds in the minerals that make up the rock were disrupted by the water molecules causing the rock to partially melt. The magma produced from this melting was more buoyant, so it rose in dikes or other molten intrusions back toward the Earth's surface where it encountered granitic rock in the continental plate. Heat from the basaltic magma partially melted the

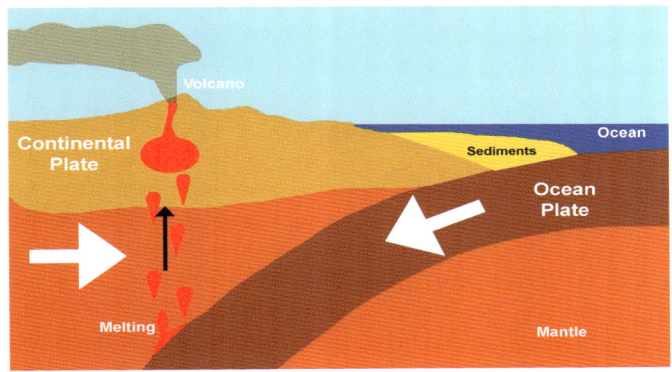

Figure 2.67: Diorite formed in plate tectonic subduction zones where ocean plates slid under continental plates.

granitic rock to form a mixed magma that was intermediate in composition between basalt and granite. If this mixed magma cooled below the Earth's surface, it crystallized into diorite. If it continued rising and erupted onto the Earth's surface, it formed extrusive igneous andesite (see Figure 2.68). The only difference between diorite and andesite is the size of the mineral grains: those in diorite are larger since they had more time to grow and develop under the Earth's surface. Given andesite's similarity to other igneous rocks, it is not featured in this book but is described on page 82.

Figure 2.68: Andesite is an extrusive igneous equivalent to diorite. Notice the smaller size of the crystal grains.

Although the mixed-mineral composition of diorite makes it less suitable for lapidary work, it has been used historically because it was easy to carve inscriptions into this rock. The use of diorite in art was most important in ancient cultures such as Egypt and Babylonia (see Figure 2.69). It was so valued 3,000 years ago that some empire rulers sent their militaries to search for and mine diorite deposits.

Figure 2.69: This Egyptian jar was carved around 1450 BC from diorite and trimmed with gold leaf.

Rock Identification: Diorite

A specimen can be identified as diorite if it has the following characteristics.

Rock Identification Tips	Photographs or other identification information
The edges of the crystals in diorite interlock because the minerals are bound on the molecular level and fit together like a jigsaw puzzle.	**Figure 2.70:** This is a cross-polarized microscopic image of diorite. The field of view is .06 inches (.15 cm). The minerals are plagioclase, hornblende, and magnetite. 
Diorite has a speckled arrangement of crystals like granite, but it is usually darker in color due to it having less light-colored quartz and more darker plagioclase feldspar, amphibole, or pyroxene crystals.	**Figure 2.71:** Diorite tends to be darker than granite. 
The mineral crystals in diorite are organized into a black and white, salt and pepper pattern.	**Figure 2.72:** Diorite has a salt and pepper (light and dark) crystal grain arrangement. 

Diorite has a hardness of between 5.5 and 6.0 on the Mohs Scale and will scratch glass.	*Figure 2.73:* Diorite will scratch glass. 
Diorite has visible crystal grains all about the same size (larger than a grain of rice) that are evenly distributed throughout the specimen.	*Figure 2.74:* This diorite shows large, randomly arranged crystal grains. 
Many of the crystal faces in diorite are flat and shiny, so they reflect light when rotated in either sunlight or a bright light.	*Figure 2.75:* Many diorite specimens have flat crystal surfaces that reflect light. 

Diorite has less than 10 percent potassium feldspar, so there should be little or no pink color.	***Figure 2.76:*** *Diorite contains almost no pink potassium feldspar.* 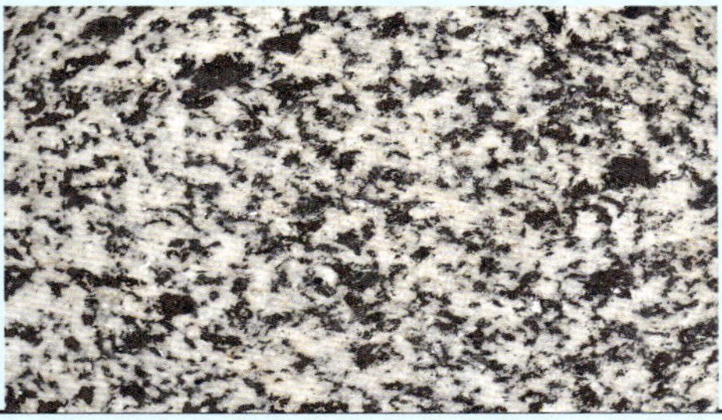
Diorite has very little light-colored muscovite mica but can contain dark biotite mica.	***Figure 2.77:*** *Diorite often has dark biotite mica.* 
Diorite can be differentiated from granite by using a hand lens. Granite has a lot of gray or clear irregularly shaped quartz grains. Less than five percent of the crystal grains in diorite are quartz. The white crystals in diorite are usually feldspar.	***Figure 2.78:*** *Diorite has almost no quartz.* 
Diorite can be differentiated from gabbro by its lighter color.	*Diorite is lighter in color than gabbro and has a salt and pepper arrangement of crystal grains, which gabbro does not. Gabbro is featured later in this chapter.*
Diorite is opaque.	*Use a bright flashlight to verify that light does not shine through the specimen.*
Diorite does not contain fossils.	*Because diorite formed below the Earth's surface from semi-molten magma, this rock does not contain fossils.*

Diorite Photographs

Figure 2.79: *A typical salt and pepper colored diorite.*

Figure 2.80: *This diorite specimen is from the Gitche Gumee Museum's collection.*

Figure 2.81: *This diorite has an epidote-filled fracture (greenish secondary fill).*

Figure 2.82: *A diorite from a Lake Superior beach.*

Figure 2.83: *Another diorite found on a lakeshore beach.*

Figure 2.84: *This diorite has a lot of hornblende.*

Figure 2.85: A typical diorite specimen.

Figure 2.86: This diorite has smaller crystal grains.

Figure 2.87: Another typical diorite.

Figure 2.88: A rare orbicular diorite.

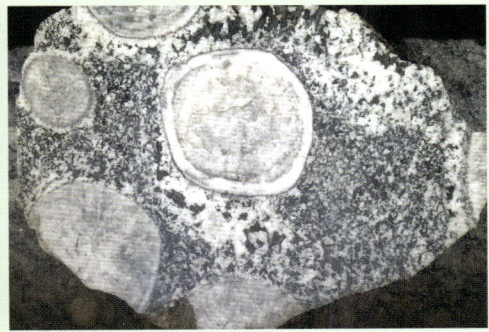

Figure 2.89: This cliff has rare orbicular diorite.

Figure 2.90: This diorite cliff is in Antarctica.

Figure 2.91: A diorite cliff with naturally faulted (split) slabs.

Figure 2.92: A close-up of diorite crystal grains.

Gabbro

Another common coarse-grained intrusive igneous rock is gabbro. This rock has large crystals easily seen without magnification. Many of the crystal faces are quite reflective, which can give gabbro a "sparkly" appearance when a specimen is rotated either in sunlight or a bright light. This is because some of the well-developed feldspar crystals are rectangular with flat sides that easily reflect light. Most of the time mineral grains in gabbro are .04 inches or larger (>1 mm) but can be up to .40 inches (1 cm). Diabase is a variety of gabbro with the same mineral composition but with smaller crystal grains (see Figures 2.93 and 2.94). Because gabbro and diabase are so similar, the latter is not featured in this book.

Figure 2.93: Gabbro.

Figure 2.94: Diabase.

Gabbro is usually black to dark gray and is darker in color than other intrusive igneous rocks. Gabbro can be tinted green if it contains epidote or olivine minerals. A minor amount of light-colored mineral grains may also be present. The dark color comes from pyroxene and plagioclase feldspar, with minor amounts of other minerals. Gabbro has a variable hardness, but it is about 5.0 to 6.0 on the Mohs scale and sometimes scratches glass but can always be scratched with a knife blade.

Gabbro formed from the same kind of magma as extrusive igneous basalt, but because it cooled more slowly, it developed larger crystals. Some of the feldspars are so well developed that they appear as large rectangular, striated (grooved) crystals. Like other igneous rocks, gabbro is used in the construction industry for everything from crushed stone to polished counter tops and floor tiles, where it is often referred to as "black granite."

A lot of gabbro formed at mid-ocean ridges, which are seafloor mountain systems running through the middle of our planet's large oceans that were (and still are) a key player in plate tectonic forces. Along these ridge systems, two continental plates moved apart when seafloor spreading was facilitated by upwelling lava (see Figure 2.95). It is often stated that Earth's oceanic crust is made of basalt. The word "basalt" is used because the oceanic crust rocks formed at mid-ocean ridges have a "basaltic" composition. However, basalt only coated the surface of oceanic crust where it cooled quickly. Underneath the basaltic coating, gabbro formed most of the oceanic crust. At greater depth, the cooling rate was slower, allowing larger crystals time to develop, so gabbro formed instead of basalt.

Figure 2.95: Gabbro often formed at mid-ocean ridges where there was (and still is) plate tectonic seafloor spreading.

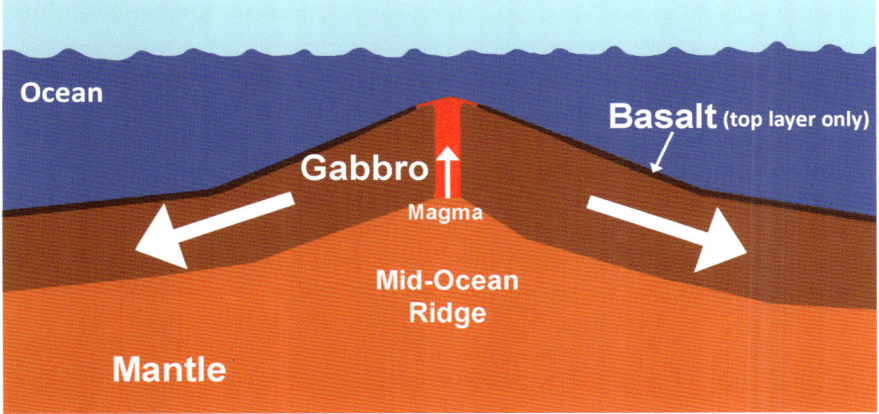

Like diorite, gabbro also developed in igneous intrusions below continental subduction zones, especially in the magma chambers that fed basaltic eruptions. Still another area where gabbro formed was in the middle of continents where thick, extensive flood basalts erupted out of rift zones (or filled rift valleys). Although basalt formed along the top of individual lava flows, inside the thicker lava flows the rate of crystallization was slower, allowing gabbro to form instead of basalt.

Gabbro, like other intrusive igneous rocks, has an interlocking granitic texture. Since texture depends upon rate of cooling, rocks of gabbro composition vary from fine-grained (diabase) that formed on the outside of a dike or sill intrusion to a typical gabbro that formed in the middle of a magma chamber or large lava flow. The mineral crystals in gabbro are deeply intermeshed, making it a very tough rock.

When uplifted to the Earth's surface by plate tectonic forces, exposed gabbro became weathered. Oftentimes the exposed surface of gabbro developed either a white bleached or brown rind which can obscure the rock's texture and composition. Thus, when attempting to identify gabbro, it is important to look at freshly broken surfaces of the specimen.

A large gabbro deposit now exposed at the Earth's surface can be found in the Duluth Complex, located west and north of Lake Superior. This geologic complex is a ten-mile-thick intrusive sill (16 km) that has been uplifted and exposed. It is composed mostly of gabbro with lesser amounts of other igneous rocks (see Figure 2.96).

Figure 2.96: *The Duluth Complex assemblage of igneous rocks contains significant amount of gabbro.*

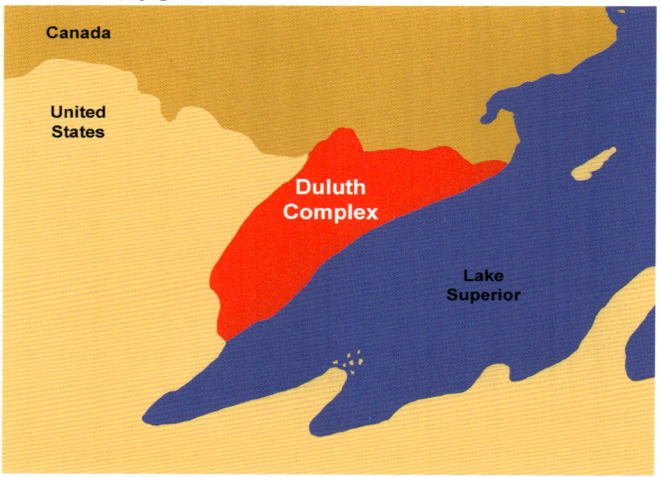

Rock Identification: Gabbro

A specimen can be identified as gabbro if it has the following characteristics.

Rock Identification Tips	Photographs or other identification information
The crystal grains in gabbro are intermeshed and bound together on the molecular level. Gabbro tends to be a very tough rock.	*Figure 2.97:* *This is a cross-polarized microscopic image of a Scottish gabbro specimen. The field of view is .79 x 1.6 inches (2 x 4 cm).*

Gabbro is usually dark gray, dark green, or black. It is almost always darker in color than either granite or diorite.	*Figure 2.98:* This gabbro is from a Lake Superior beach. It most likely originated from the Duluth Complex. 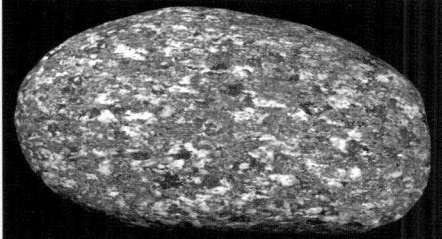
Gabbro is mostly dark in color but can have some lighter crystal grains.	*Figure 2.99:* Gabbro is mostly dark but can have light-colored crystal grains. 
Gabbro has dark feldspar crystal surfaces that are shiny when the specimen is rotated in either sunlight or a bright light. Oftentimes these crystals are rectangular in shape.	*Figure 2.100:* This gabbro has large feldspar crystals, as well as some quartz. Quartz-bearing gabbro specimens are rare. 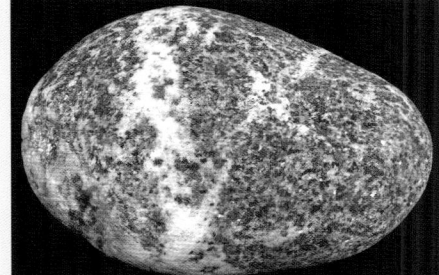
Gabbro is coarse-grained: most of the crystal grains are larger than a grain of rice. The "first cousin" diabase has crystal grains almost always smaller than a grain of rice.	*Figure 2.101:* Visible gabbro crystals are usually larger than a grain of rice. 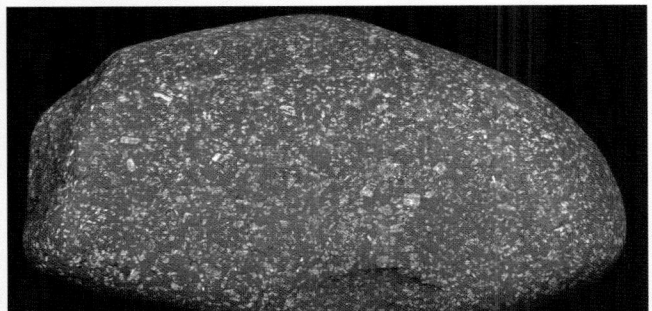

If there are lighter colored feldspar crystals, they usually appear "fuzzy" or "out-of-focus" because they are not well defined. After forming first, these light-colored crystals were squeezed by darker minerals that formed around them and distorted their shape.	*Figure 2.102:* Light colored crystal grains in gabbro can appear fuzzy when they become engulfed by darker pyroxene crystals that formed later in the cooling and solidification process. 
When exposed to weather, the outer surface of gabbro often lightens and/or turns brown.	*Figure 2.103:* Weathered gabbro specimens can have a brown-colored husk. 
Gabbro has a hardness that varies between 5.0 to 6.0 on the Mohs Scale, so it sometimes will scratch glass. A knife blade will always scratch gabbro.	*Figure 2.104:* Gabbro specimens sometimes will scratch glass. 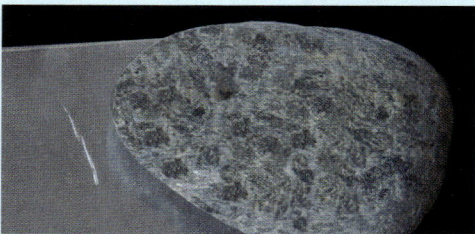
Although the mineral grains in gabbro tend to be similar in size, specimens can be porphyritic and contain larger crystals.	*Figure 2.105:* This porphyritic gabbro specimen has some large crystals. 

Gabbro contains iron bearing minerals, so a specimen often seems heavier than other rocks the same size.	**Figure 2.106:** *This specimen seems heavy because it contains a lot of iron bearing minerals. It grades into diabase due to the smaller crystal grains.*
Because gabbro usually does not contain potassium feldspar, it has no pink coloration. Gabbro does contain plagioclase feldspar.	*The lack of pink coloration in gabbro differentiates it from many granite specimens.*
Gabbro can be differentiated from granite by its lack of quartz.	*Gabbro is significantly darker in color than granite because of its lack of quartz.*
Gabbro is opaque.	*Use a bright flashlight to check for translucency (or lack of).*
Gabbro specimens never contain fossils.	*Because gabbro formed below the Earth's surface from semi-molten magma, it does not contain fossils.*

Gabbro Photographs

Figure 2.107: *This gabbro has green olivine crystal grains.*

Figure 2.108: *This gabbro is from a Lake Superior beach.*

Figure 2.109: This specimen is halfway between gabbro and diabase.

Figure 2.110: This gabbro specimen has more light colored feldspar than usual.

Figure 2.111: A gabbro from a Lake Superior Beach.

Figure 2.112: This gabbro has large light-colored feldspar crystals and green olivine crystals.

Figure 2.113: This gabbro is from South Africa.

Figure 2.114: A weathered gabbro specimen.

Figure 2.115: This gabbro is from Sweden.

Figure 2.116: A gabbro from a Lake Superior beach.

Figure 2.117: This specimen is an intermediate between aabbro and diabase.

Figure 2.118: A close-up of diabase.

Figure 2.119: This gabbro is from Russia.

Figure 2.120: Another close-up of diabase.

Figure 2.121: Antarctica gabbro.

Figure 2.122: This Minnesota gabbro shows layers that resulted from repeated episodes of crystal settling in a magma chamber.

53

Syenite

Syenite is a medium- to coarse-grained intrusive igneous rock similar to granite but with little or no quartz and almost no mica. Granite has between 20 and 50 percent quartz, whereas syenite usually has less than five percent quartz by volume. Syenite is feldspar dominant with over 65 percent feldspar. Of the feldspars in syenite, around 90 percent is potassium and ten percent plagioclase. Sometimes the feldspar types became intergrown during crystal formation. Like granite, syenite is defined by its mineral composition. The color of syenite can vary significantly, but it is usually lighter in color and can be pink, light red, white, light gray, pale green, or pale brown. It has a hardness of between 5.5 and 6.0, so it will scratch glass.

Syenite formed within continental rift zones and subduction zones. For syenite to develop, there had to be existing first-generation granitic igneous rock that was exposed to hot magma, which resulted in a low degree of partial melting. Higher degrees of melting would have formed granite or other intrusive igneous rocks. At very low degrees of partial melting, the magma contained lower concentrations of silica, which developed into syenite when the melt completed its cooling and crystallization process.

In recent years, a certain variety of syenite has been discovered on the shores of the Great Lakes. These syenite pebbles contain fluorescent sodalite (see Figure 2.123). Sodalite is a scarce mineral that formed in silica-poor rocks, which in and of itself is rare because silica is one of the most common minerals in the Earth's crust. Most people are familiar with blue sodalite (see Figure 2.124), but sodalite can be white, gray, purple or green. The sodalite in fluorescent syenite is white or gray so under normal light it is difficult to distinguish from the other component minerals. Most syenite in the world is not fluorescent, including large syenite deposits located north of Lake Superior. But the variety of syenite featured in this chapter, which originates east of Lake Superior in Ontario, does contain varying amounts of fluorescent sodalite. It appears, however, this is not the only fluorescent syenite deposit. Once word spread on the Internet, people in different geologic locations have been searching for fluorescent varieties of syenite. For example, a woman in Massachusetts has found fluorescent syenite in her state.

Figure 2.123: Syenite with fluorescent sodalite.

Figure 2.124: Blue sodalite that is not fluorescent.

Fluorescence is detected by using ultraviolet (UV) light, which is invisible light with wavelengths between 100 and 395 nanometers (nm). These wavelengths are shorter than the 380 to 700 nm range that humans can see. Fluorescence occurs when one wavelength outside of the visible spectrum is absorbed by a mineral and re-emitted at a longer visible wavelength. This occurs at the molecular level when electrons in atoms get excited by the UV light and jump away from the nucleus to an outer electron shell. Another electron fills that spot and releases energy in the form of a visible glow. Different minerals will fluoresce with different wavelengths of UV light. In the case of the fluorescent sodalite in this variety of syenite, the mineral fluoresces yellow-orange (if the UV light is filtered), or pink and purple with an unfiltered UV light (because the visible purple light is not filtered out). The peak excitation wavelength for this sodalite is at a longwave UV wavelength of 385 nm, although it will fluoresce at wavelengths between 365 and 395 nm. Most of the flashlights available on the market emit a 365 nm wavelength.

After scientific investigation, it appears this rare variety of fluorescent sodalite, called hackmanite (see Figure 2.125), has a slight difference in its chemistry. Within its chemical structure, some of the chlorine has been replaced by sulfur. Sodalite without this chemical difference does not fluoresce. After this variety of sodalite was first found on a Lake Superior beach in Michigan's Upper Peninsula, it has been nicknamed "Yooperlite," referring to people who live in the Upper Peninsula of Michigan (Yoopers). Fluorescent rock collectors, however, have known about this fluorescent variety in Canada for many years.

Figure 2.125: This is a fluorescent hackmanite specimen (a variety of sodalite) from Afghanistan.

It is interesting that most fluorescent rocks are activated by shortwave fluorescent wavelengths around 254 nm. The 365 nm wavelength was originally developed for use in industrial inspections to find cracks and leaks, to cure adhesives and resins, and to find animal urine and blood splatter. Since the discovery of fluorescent syenite on the Great Lakes beaches, many more longwave UV lights have become available on the market. Some people have attempted to find fluorescent syenite during daytime hours. This is almost impossible since syenite rocks look like other granitic varieties. Without a UV light with the right wavelength, and the darkness needed to see its effect, it is difficult to see fluorescent syenite during the day. The two photos below (see Figures 2.126 and 2.127) show the same part of the same syenite specimen: one under 365 nm light and one under normal light. Note the fluorescent sodalite mineral grains are not visible under normal light.

Figure 2.126: This syenite specimen fluoresces under 365 nm UV wavelength (field of view = 3 inches wide/7.6 cm).

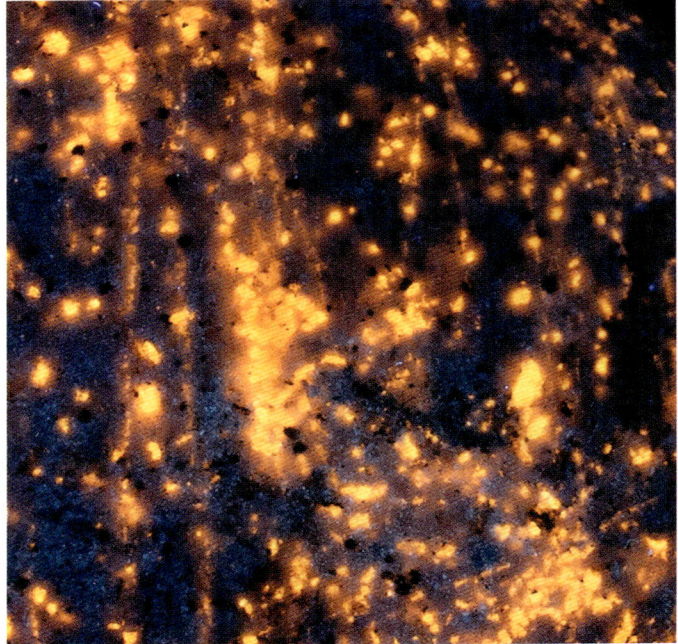

Figure 2.127: Here is the same specimen of fluorescent syenite under normal light.

Although fluorescent syenite can now be found throughout the upper Midwest, it originated in central Ontario and was dragged by glaciers to the southwest during the last period of glaciation which ended around 10,000 years ago. Even before the fluorescent property of this syenite was discovered, it was not a common rock on Great Lakes beaches. Since its discovery, people have flocked to the area in search of this fluorescent rock, which has resulted in it becoming harder and harder to find. Once all the fluorescent syenite specimens are picked up from the Great Lakes beaches, the supply will be gone and cannot be replenished — at least not until the next glacier scrapes it from the source in central Ontario and deposits more across the landscape. So PLEASE, if you go out to find these amazing rocks, just take a few and leave the rest for others to find.

Rock Identification: Fluorescent Syenite

Because this chapter is only featuring fluorescent syenite, all that is needed to identify a specimen is to acquire a UV flashlight. The best light to use emits a 365 nm wavelength and has a built-in filter that removes light in the visible spectrum. Many of the first UV lights available to rockhounds were not very powerful and did not have a filter. In the last few years, several UV flashlights have been introduced to the market due to the demand of fluorescent syenite hunters. For those interested in purchasing a UV light, keep in mind that not all 365 nm flashlights are created equally. Several factors make some UV lights better than others including: the number of LEDs, the strength of the total LED output, the shape and quality of the reflector inside the flashlight, and battery strength. When comparing flashlights, the most important factor is the useful distance of the flashlight. This is significant because sometimes it is hard to spot fluorescent syenites that contain a lesser amount of sodalite. The proportion of fluorescent sodalite versus the nonfluorescent portions of syenite specimens varies significantly. The cheaper lights will only shine a beam two to three feet, whereas the more expensive UV flashlights have a usable beam that shines 6 to 12 feet (2 to 4 m). One gauge as to how effective a UV flashlight will be in finding fluorescent syenite is price. The cheaper flashlights ($20 to $40) often do not have a beam strong enough to reach the rocks on the ground when walking the beach. Studies have shown that a weaker flashlight will only have 25 percent or less the intensity of a stronger UV flashlight at the same distance. If you have a lesser expensive UV light, it can help to tape the flashlight to a stick to move it closer to the ground, but the size of the light beam will limit the number of rocks you can see at a time. A stronger UV flashlight will have a wider beam, will reach farther, and will illuminate more rocks. To increase the chances of success in searching for fluorescent syenite, it is best to spend at least $100 on a good quality UV flashlight.

Syenite Photographs

Figure 2.128: A cross-polarized microscopic photograph of syenite.

Figure 2.129: Two specimens of fluorescent syenite.

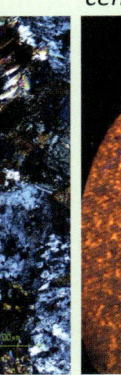

Figure 2.130: Syenite fluoresces under 365 nm UV light.

Figure 2.131: Here are the same specimens as Figure 2.130 under normal light.

Figure 2.132: Fluorescent syenite under 365 nm UV light.

Figure 2.133: The same specimen under normal light.

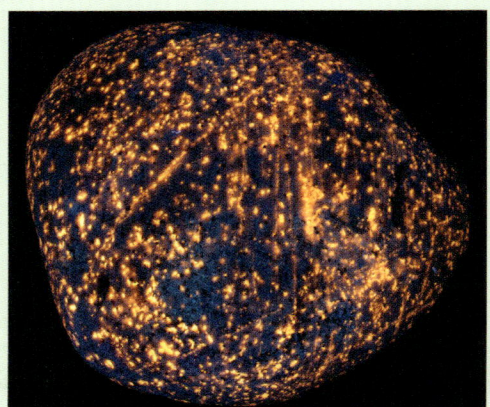

Figure 2.134: *This nonfluorescent Nepheline Syenite is from Wisconsin.*

Figure 2.135: *Here is another fluorescent syenite from a Lake Superior beach.*

Figure 2.136: *Here is another nonfluorescent Nepheline Syenite. Like granite, syenite has many different appearances.*

Figure 2.137 *A nonfluorescent syenite from Vermont.*

Chapter 3: Extrusive Igneous Rocks

Note: For a list of the Featured Rocks included in this chapter, please go to page 67.

Geologic Background

To tell the story of how extrusive igneous rocks formed in the upper Midwest of the United States, it is necessary to review the area's geologic history. Each geographic region has its own geologic story, but this one is unique and interesting. Beginning around 4.3 billion years ago, volcanic islands and ocean crust had already developed in the area that is now North America. Four billion years ago, partial melting of the lower portion of the basaltic ocean crust began. This melting created lighter, silica-rich intrusive rock that "floated" on top of the Earth's mantle and became the first large stable landmasses on Earth, called cratons. During Precambrian time (4.6 billion to 544 million years ago), plate tectonic movement of these landmasses caused many collisions that joined the granitic cratons together. By 2.7 billion years ago, the largest craton on our planet, the Superior Craton, had taken shape (see Figure 3.1).

As convection currents in the mantle (see Figure 3.2) continued to fuel plate tectonic forces, the Superior craton was propelled

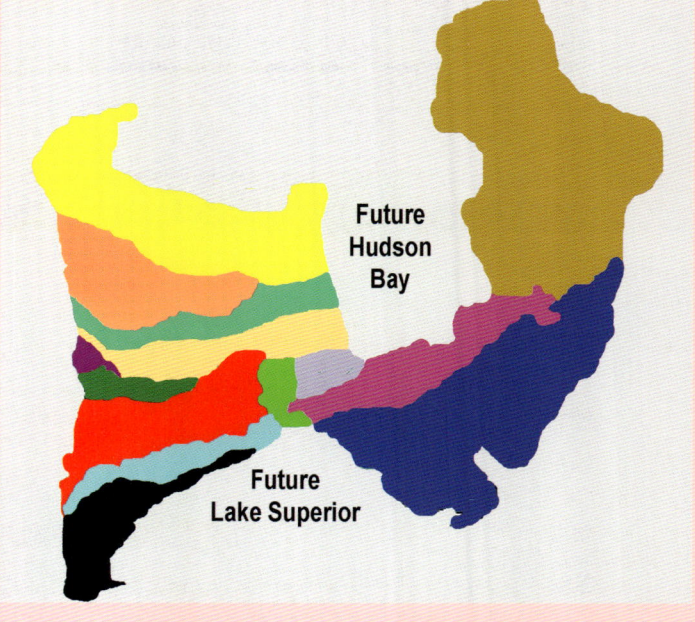

Figure 3.1: Superior Craton — the original core of the North American continent. This graphic shows how small landmasses combined to form the Superior Craton.

around the globe. It collided with additional cratons, further increasing the size of the landmass that grew into the Canadian Shield. Craton collisions were especially common between 1.0 and 1.5 billion years ago. The Canadian shield increased in size to become the continent of Laurentia, which later evolved into the North American continent (see Figure 3.3).

Figure 3.2: Convection currents in the Earth's mantle push tectonic plates around the globe.

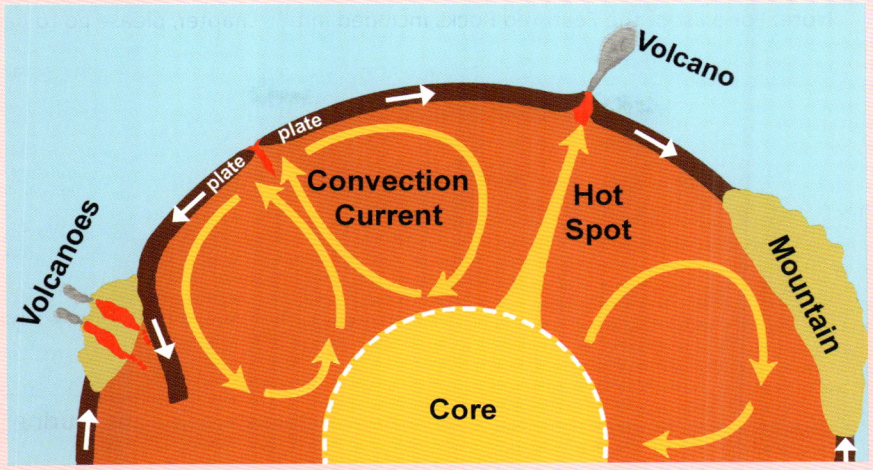

Figure 3.3: Assembly of the North American continent. The light blue areas (mostly sedimentary rocks) were added later in geologic time.

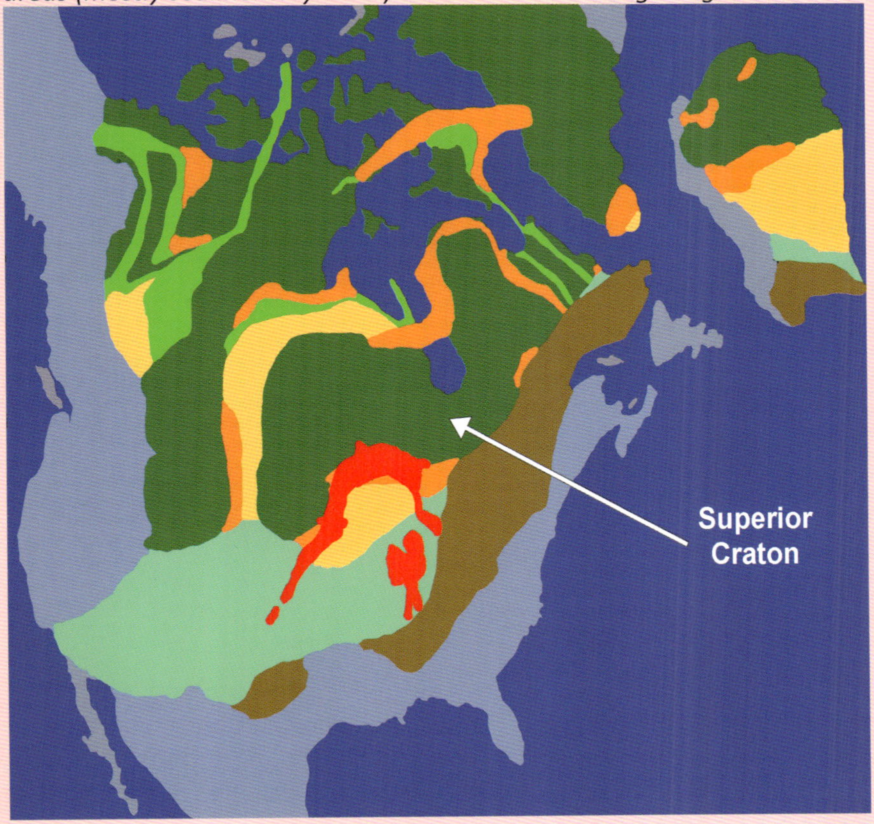

As plate tectonic forces continued, most of the landmasses existing 1.1 billion years ago collided to form the supercontinent, Rodinia (see Figure 3.4). Although the geologic record for this period is sparse, it is thought that Rodinia contained about 75 percent of the continental landmass present today. There is geologic evidence, however, indicating that the collision of Laurentia with other cratons during the assembly of Rodinia caused long mountain ranges to form, including the Grenville Provence (see Figure 3.6). When Rodinia broke up between 633 and 750 million years ago, the Laurentian continent separated from other cratons, including Amazonia (see Figure 3.5).

Figure 3.4: *The Superior Craton grew to be the Canadian Shield, which then enlarged to be the Laurentia continent. Laurentia combined with all the other land masses on Earth to become the supercontinent of Rodinia around a billion years ago.*

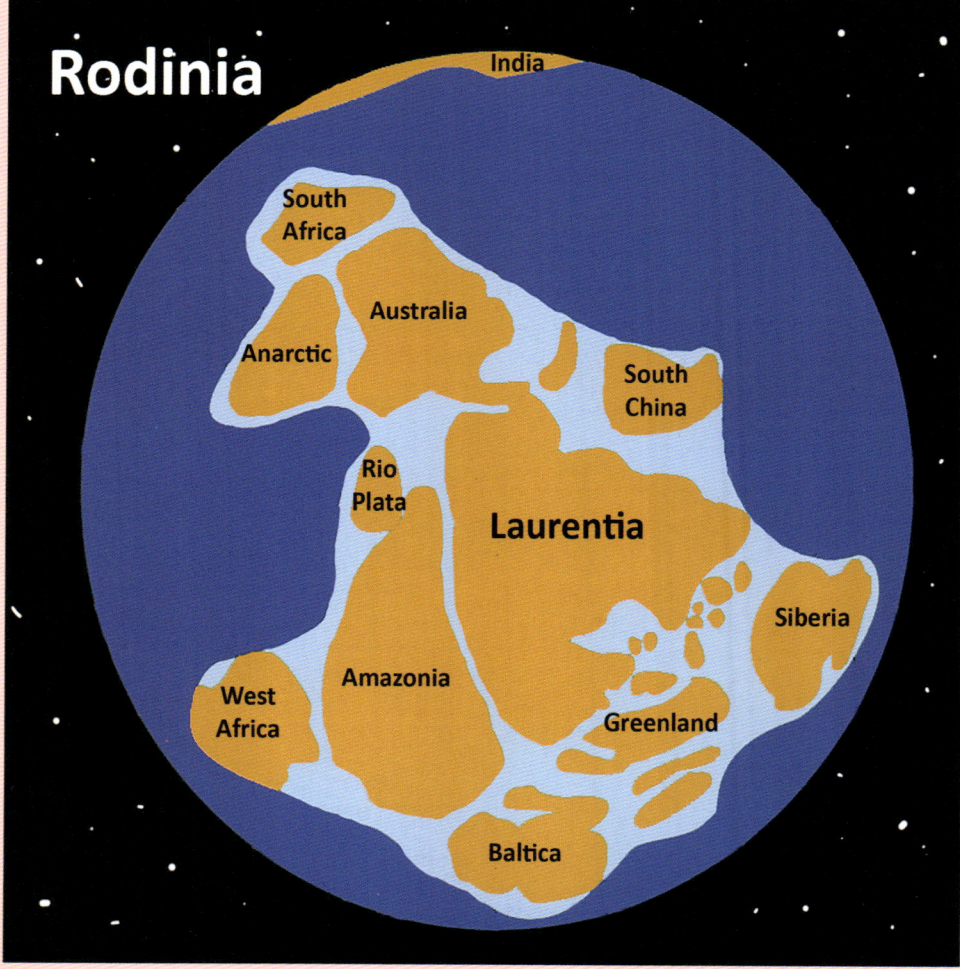

During the time of the Laurentian continent, two major incidents happened simultaneously to create a geologic wonder in the area that is now the Lake Superior region. These events were the Midcontinent Rift and the development of a hot spot mantle plume under the rift (see Figure 3.6). Mantle plumes are driven by heat exchange and convection across the core-mantle boundary located nearly 1,900 miles (3,000 km) below the Earth's surface (see Figure 3.7). The heat causing this convection was from the continued decay of radioactive elements deep inside the planet. Today Earth still has enough heat for active convection that will continue for a long time. In comparison, Planet Mars which is much smaller— about the size of Earth's core — has significantly cooled and no longer has plate tectonic activity.

Figure 3.5: The Rodinia supercontinent split apart between 633 and 750 million years ago.

Figure 3.7: A mantle plume pushed semi-molten magma toward the Earth's surface. which helped trigger the Midcontinent rifting event.

Figure 3.6: A mantle plume hot spot developed under what would become the Midcontinent Rift.

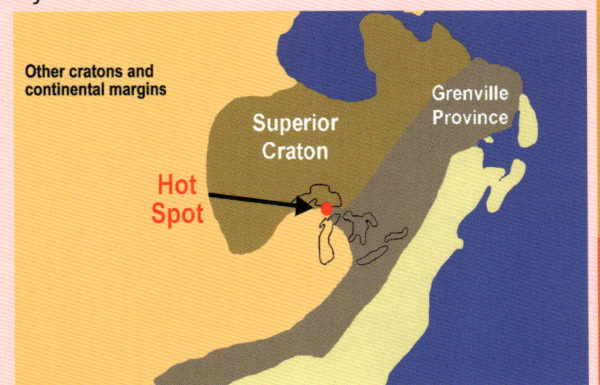

In the area that is now Lake Superior, convection currents moved hot, buoyant magma toward the surface. This hot spot had a thin rising column connecting to a bulbous head that expanded in size as the plume rose. Magma pushed up through this hot spot plume with such force that

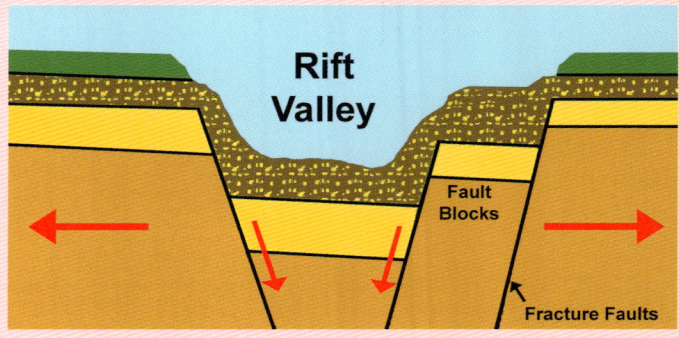

Figure 3.8: In what is now the Lake Superior region, a rift valley formed where the Earth's crust cracked above the mantle plume and began moving apart.

it domed up and fractured the Earth's crust. The doming produced large-scale extension that caused the continental crust to stretch and break along newly formed faults. Earthquakes occurred along these faults, which moved huge blocks of rock downward on both sides of the cracked crust. The landmasses on either side moved away from each other causing a rift valley to form in the middle (see Figure 3.8). This process can be visualized by pulling both ends of a candy bar in opposite directions. The outer chocolate coating breaks while the inside filling stretches and bends downward.

The Earth's crust continued cracking away from the rift valley, forming two "arms" of the rift that extended to the southeast and southwest (see Figure 3.9). It has been difficult for scientists to determine the exact dimensions of the rift because most of it is now buried under sediments deposited over the last billion years. In fact, scientists did not discover the existence of this hot spot plume and rift valley until the 1960s. With recent imaging technologies, it has been determined that the southeast arm of the rift went through what is now central lower Michigan and

Figure 3.9: Two arms of the Midcontinent Rift cracked the Earth's crust away from the rift valley.

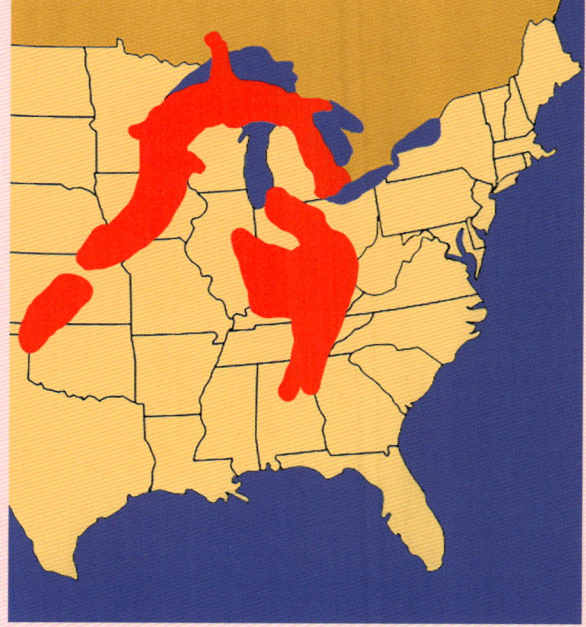

continued through parts of Indiana, Ohio, Kentucky, Tennessee, and Alabama. The southwest arm went through portions of Wisconsin, Minnesota, Iowa, Nebraska, Kansas, and Oklahoma. A third northern arm of what is called a triple junction failed and did not develop. Scientists estimate the rift averaged a hundred miles wide and was almost 2,000 miles long (160 x 3,200 km).

Not only did the Earth's crust split apart during this rifting event, but the crack in the Earth's crust created volcanoes in what is now western Lake Superior. These volcanoes erupted and spewed out lava and ash. In addition, at other spots along the rift, tremendous amounts of lava flowed out of vents and filled the rift valley (see Figure 3.10).

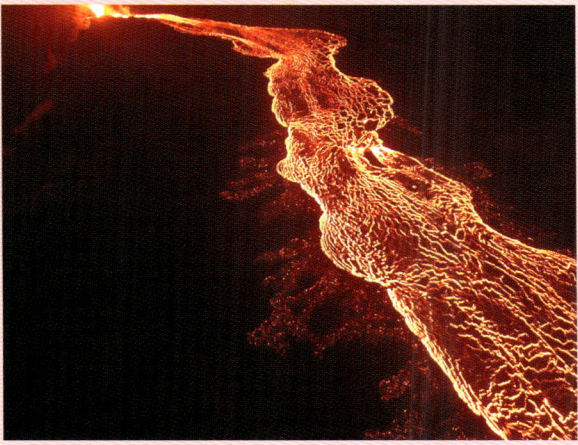

Figure 3.10: Lava flowed from both volcanoes along the rift and the mantle hot spot plume and filled the rift valley.

There are several reasons why the Midcontinent Rift was unique in Earth's geologic history. First, in most cases when a rifting event happens, the rifting continues until the landmass splits in two. In the case of the Midcontinent Rift, that did not happen. If it had occurred, the North American continent would have split in half and an ocean would have formed between the two halves. Second, other rift valleys fill with mostly sedimentary rock. The Midcontinent Rift valley instead initially filled mostly with lava, although sedimentary layers were deposited between lava flows and after the igneous activity ended. Third, the amount of lava that poured into the rift valley was beyond what would have been expected. Apparently, even after the rifting stopped, additional lava poured into the rift basin. The larger than expected amount of lava was supplied by the hot spot plume that fed lava into the rift valley.

Scientists are still investigating why the Midcontinent rifting stopped. The predominate idea is that the Grenville province landmass (see Figure 3.6) continued moving to the northwest toward what is now the Great Lakes region, which compressed the rift and closed it shut. When the Midcontinent Rift failed, the faults that formed during the rifting event reversed direction. This reversal in crustal movement along the rift compressed the rock layers and formed a U-shaped syncline through what is now the center of Lake Superior. Synclines are folds in which each half of the fold dips toward the center channel of the fold (see Figure 3.11). The volcanic rocks and associated sediments on the north side — as seen at Isle

Royale — dipped southward toward the center of Lake Superior, whereas those on the south side — as seen in the Keweenaw Peninsula — dipped to the north. The reverse compression within the syncline uplifted and exposed volcanic rock on both sides of the syncline (see Figure 3.12).

Figure 3.11: When Midcontinent rifting stopped, crustal movement reversed direction, which caused a syncline to form through the middle of what is now Lake Superior.

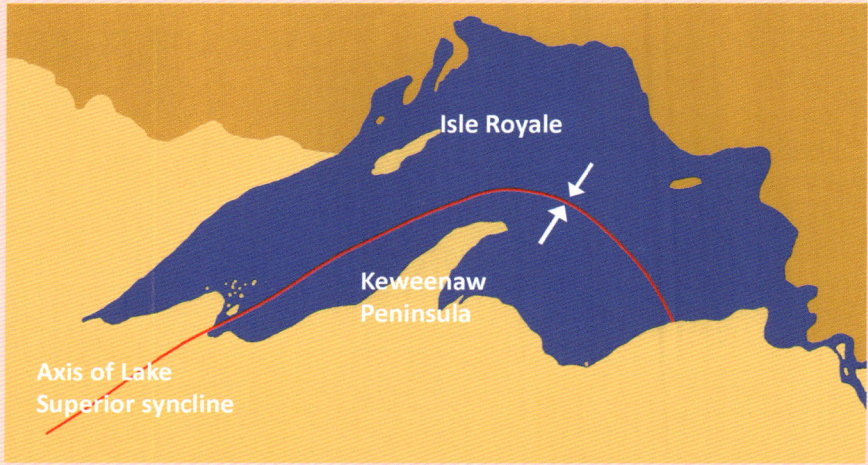

Figure 3.12: As geologic forces pushed the sides of the rift valley back toward each other, the compression uplifted and exposed volcanic rock on both sides of the syncline. This diagram shows the profile view.

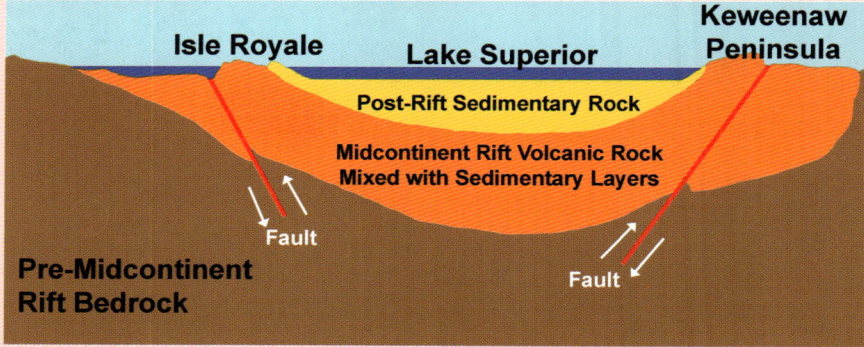

A significant impact the Midcontinent Rift had on the upper Midwest was the amount of igneous rock that was produced. Since most of the lava that flowed or erupted from the Midcontinent Rift cooled and crystallized at the Earth's surface to form extrusive igneous rock. Exposure to the relatively cool temperatures at the planet's surface made the lava solidify quickly, over a period of minutes to months in most cases. Rapid cooling allowed the individual mineral grains only a short time to grow, so their final size was tiny and fine-grained (less than .02 inch, ½ mm). To see the minute crystal grains in extrusive igneous rocks, magnification is usually required.

There are several types of extrusive igneous rocks that formed in the Midcontinent Rift, as well as at other places on Earth where there was volcanic activity. Extrusive igneous rocks formed in two ways: (1) from lava flows, and (2) from the volcanic ejection of fragmented pieces of rock and ash (pyroclastic materials) into the atmosphere that settled out and blanketed the Earth's surface. The small grain size in extrusive igneous rocks makes accurate identification of specimens a challenge without the use of sophisticated laboratory equipment, which is not available to most rockhounds. Thus, field identification of extrusive igneous rocks relies on the rock's color, texture, and any visible phenocrysts (large crystals embedded in a fine-grained rock).

Featured Rocks

Although there are many different types of extrusive igneous rocks, this book will feature two main types, describe six other extrusive igneous rocks, and mention 11 associated amygdaloidal minerals.

> **Key**
> - Featured rocks (includes rock identification information and photographs).
> - Rocks are described but not featured (includes photographs).
> - Minerals are described (includes photographs).
> * Varieties of these rocks from the Great Lakes region are included as examples.

- **Basalt** (page 68): A dark-colored extrusive rock with a fine-grained texture that formed from volcanic lava. Different forms of basalt are described including:

    - Vesicular Basalt (page 72): Basalt with open, unfilled pockets that formed from gas bubbles in the lava;

    - Scoria (page 73): a form of vesicular extrusive igneous rock;

    - Porphyritic Basalt (page 74): Basalt with large crystals embedded in a small-crystal groundmass;

    - Amygdaloidal Basalt (page 75): Basalt with vesicle pockets that filled in with minerals after the basalt solidified.

    NOTE: Examples of amygdaloidal minerals that fill the vesicle pockets are at the end of this chapter, beginning on page 93.

- **Andesite** (page 82): This is an intermediate extrusive rock between basalt and rhyolite.

- **Rhyolite** (page 85): A silica-rich extrusive igneous rock that formed from lava or pyroclastic materials.

- **Pumice** (page 92): A vesicular extrusive igneous rock that formed from explosive eruptions.

- **Amygdaloidal Minerals** (pages 93 to 124):
    - Chalcedony
    - Agate
    - Carnelian
    - Calcite
    - Chlorastrolite (Isle Royale Greenstone)*
    - Copper
    - Datolite*
    - Epidote
    - Milky quartz
    - Prehnite*
    - Thomsonite*

Basalt

Most of the igneous rock formed during the Midcontinent rifting event was basalt. Basalt makes up 90 percent of extrusive rock on our planet. It is composed mostly of gray plagioclase feldspar and dark colored pyroxene but can have minor amounts of olivine, quartz, or hornblende (see Figures 3.13 to 3.17). Basalt formed from lava with a temperature range between 1,832°F to 2,192°F (1,000°C to 1,200°C). Like other extrusive igneous rocks, basalt cooled and crystallized at the Earth's surface. As lava reached the planet's surface, it had four parts: a hot liquid base, called the melt; intact mineral crystals with high melting temperatures; solid pieces of rock incorporated into the melt that broke off from rock surrounding the magma chamber; and dissolved gases. Deep below the surface, when rocks melted to initially form magma, they did so slowly and gradually. When there was enough heat to melt the minerals in the rock that had the lowest melting temperatures, tiny globules of molten melt formed between mineral grains, which linked them together, softened the rock, and generated more melting. The once solid rock turned into magma when between 20 and 35 percent of the solid mass melted. Rarely does magma contain more than 50 percent molten rock. Once the magma reached the Earth's surface, it cooled and crystallized into basalt.

Figure 3.13: *Feldspar.*   ***Figure 3.14:*** *Pyroxene.*   ***Figure 3.15:*** *Quartz.*

Figure 3.16: *Olivine.*   ***Figure 3.17:*** *Hornblende.*

Basalt is usually gray or black but can also be tinted brown or dark green from the presence of other minerals. Basalt is one of the most abundant igneous rock on Earth, and makes up much of the surface of Venus, Mars, and other planetary bodies. Olympus Mons, for example, is a shield volcano on Mars. It, like most other volcanic features on Mars, was formed from basaltic lava flows (see Figure 3.18). It is the highest mountain on Mars and is the largest known volcano in the Solar System at about 72,000 feet tall and over 370 miles in diameter (21,946 m, 624 km).

Figure 3.18: *Basaltic volcano on Mars, Olympus Mons. This is the largest volcano in our Solar System.*

Basaltic lava has erupted in a wide variety of tectonic environments all over our planet including along mid-ocean ridges, island arcs (curved chains of volcanic islands located along tectonic plate margins), hotspot volcanic islands, and intra-continental rifts. Basalt rocks found today on beaches or in glacial till (rocks moved by the glaciers), are usually smooth to the touch. When freshly broken, basalt has a dull surface. Although basalt specimens often appear to have a solid color, they are composed of minute mineral grains that formed when the lava quickly cooled (<.02 inch, .5mm). The mineral grains in basalt are not usually discernible to the naked eye but can be seen with either a good hand lens or microscope (see Figure 3.19 to 3.20). The grains are randomly and evenly distributed throughout the rock. The composition of basalt is similar to gabbro, which is basalt's intrusive igneous rock "cousin." The difference between basalt and gabbro is the size of the mineral grains: basalt is fine-grained while gabbro is coarse-grained.

Figure 3.19: Cross-polarized microscopic image of basalt showing small crystal grains.

The Midcontinent rifting event in the Lake Superior region occurred over a 20-million-year period, but the lava flows responsible for forming basalt occurred during just a two-million-year portion of that time. The lava flows were intermittent producing layer after layer of basalt stacked in a vertical sequence, filling the rift valley like a bathtub. Between separate lava flows, rivers carried sediments into the rift valley. There were hundreds of separate lava flows, each ranging in thickness from a few inches to hundreds of feet, plus dozens of interbedded sedimentary layers. The thinner flows cooled quickly in a manner of days or weeks while thicker flows took years to cool. Semi-molten basaltic lava had a fairly runny viscosity due to its low silica content (between 45 and 52 percent), resulting in fast moving flows. How far each lava flow traveled depended on the lava temperature, silica content, quantity of lava, flow rate, and slope of the land. Although lava from the Midcontinent Rift was

Figure 3.20: In this cross-polarized microscopic image of basalt a gypsum accessory slide was inserted above the specimen slide to enhance color and increase analytical capabilities.

up to 100,000 times more viscous (thicker) than water, it was able to easily flow and spread into the rift valley before cooling and solidifying. In most cases lava flowed around .25 miles per hour (.4 km/h) but possibly reached up to 30 miles per hour when flowing down hill (40 km/h).

By the time Midcontinent rifting ended, tremendous amounts of igneous rock had solidified in the rift valley. The weight of the giant pile of lava rock caused the Earth's crust to depress the original rift valley. Scientists have used gravity surveys to measure the thicknesses of the buried volcanic rock. The igneous rock contains iron and is denser and more magnetic than the surrounding rock, which allows the giant basaltic deposit to be mapped. By measuring the density of rocks in the syncline depression that is now under Lake Superior, scientists have determined the igneous rock to be 15 miles thick (25 km) equaling more than 240,000 cubic miles of volcanic rock (a million cubic kilometers). The height of the now buried basalt pile is almost two and a half times higher than Mount Everest, the tallest mountain on Earth today (see Figure 3.21)! The estimated volume of basaltic rock is 44 times more than the volume of water contained in all the Great Lakes combined!

Figure 3.21: *The estimated vertical height of the basalt that filled the Midcontinent Rift valley is more than twice the height of Mount Everest!*

Basalt sometimes solidified in a distinctive fashion. When basalt or other extrusive igneous rocks cooled, there was often shrinkage, or contraction, within the rock mass. In cases when basaltic lava cooled rapidly, there was differential shrinkage — the basaltic rock would shrink in the vertical dimension, but it could not accommodate shrinking in the horizontal direction without fracturing. The extensive fracture network within basaltic outcrops sometimes formed vertical columns that were hexagonal in cross-section. The size of the columns depended on the rate of cooling; very rapid cooling resulted in small columns (.4 inches/1 cm), while slow cooling produced larger and longer columns that sometimes developed into impressive geometric cliffs (see Figures 3.22 to 3.24).

Figure 3.22: Billion-year-old basaltic cliffs near Split Rock Lighthouse on the north shore of Lake Superior.

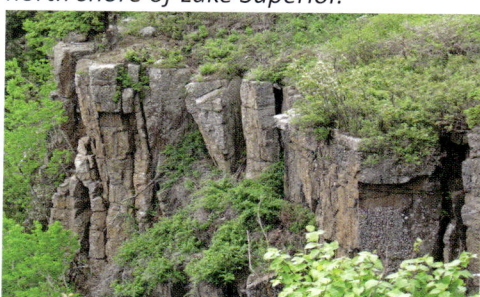

Figure 3.23: Basalt columns along the Wailuku River in Hawaii.

There is a considerable amount of basaltic rock not only in the upper Midwest but all over our planet. Depending on the lava components and the conditions impacting each lava flow, there are several varieties of basalt that developed when the lava cooled and solidified. Some of these basalt varieties are described below.

Figure 3.24: Basalt columns in Russia's Kuril Islands

Vesicular Basalt

As magma rose toward the Earth's surface, the semi-molten rock encountered lower pressure the closer it moved to the surface. The same thing happens when you open a bottle of soda pop. When the cap is removed, pressure is lessened causing carbon dioxide to come out of solution and form bubbles. In rising magma, dissolved gases came out of solution and formed bubbles. In some cases, these bubbles rose and escaped into the atmosphere. Oftentimes the bubbles became trapped in the hardening lava and served as "space savers," which formed holes or vesicular pockets (see Figure 3.25 and 3.26). Differential cooling within the solidifying lava flow resulted because the top layer exposed to the atmosphere cooled off more quickly than the lower parts of the lava flow. These temperature differences caused the hardening rock to fracture the way ice cubes crack when you put them in warm liquid. These fractures in the basalt allowed gas bubbles to escape, leaving behind a series of hollow cavities connected by a fracture network. This basaltic rock, which resembles Swiss cheese, has a vesicular texture if these cavities remained empty. Vesicles range in shape from spherical to elongated to irregular and range in size from .04 to .40 inches (1mm to 1cm).

Vesicular Basalt Photographs

Figure 3.25: Cross section of a vesicular lava tube ceiling in Utah.

Figure 3.26: Typical vesicular basalt.

Scoria

Scoria is a vesicular volcanic rock that formed from basaltic lava (see Figures 3.27 to 3.30). It differs from vesicular basalt in that it has more abundant and smaller vesicle pockets, thinner vesicle walls, and is less dense. Scoria is relatively low in density because of its numerous vesicles, but in contrast to pumice that forms from rhyolitic lava (see rhyolite section below), scoria has a specific gravity greater than one that usually makes it sink in water, compared to most pumice specimens that usually float. Scoria is generally black, dark gray, brown, dark brown, reddish-brown, dark green, or purplish red. The vesicles are spheroidal and can or cannot intersect one another.

Scoria Photographs

Figure 3.27: Typical scoria with many vesicle pockets.

Figure 3.28: Weathered scoria from Japan.

Figure 3.29: Another scoria specimen.

Figure 3.30: A close-up of scoria.

Porphyritic Basalt

Some basaltic specimens are porphyritic, which means they contain two distinctly different sizes of mineral crystals produced during two separate stages of cooling. The larger mineral crystals formed in magma at depth prior to the magma rising to the surface. These crystals formed early and stayed intact in the molten magma during transport to the Earth's surface. These large and conspicuous crystals became embedded in a fine-grained groundmass with minute crystals that cooled at or near the Earth's surface (see Figure 3.31). The larger crystals are called phenocrysts. Phenocrysts in basalt include augite pyroxene, olivine, or plagioclase feldspar. The porphyritic term is used as a modifier. For instance, a basalt with visible phenocrysts is termed a basaltic porphyry or porphyritic basalt. Other types of igneous rock also form porphyritic texture including andesite, rhyolite, gabbro, and granite.

Porphyritic Basalt Photographs

Figure 3.31: Basalt specimen with feldspar phenocrysts.

Figure 3.32: This weathered porphyritic basalt is from Utah.

Figure 3.33: Porphyritic basalt with olivine phenocrysts (green spots).

Amygdaloidal Basalt

Sometimes vesicles in extrusive igneous rocks, especially basalt, subsequently became filled with secondary minerals long after the lava cooled and solidified (see Figure 3.34). The filled cavities are called amygdules — a rock full of them is called amygdaloidal. The amygdules can be composed of chalcedony and its varieties, agate and carnelian, as well as calcite, copper, epidote, milky quartz, and other minerals. In the Lake Superior region, amygdules include chlorastrolite (Isle Royale greenstone), datolite, prehnite, and thomsonite. Some amygdules are stretched while others are round. Stretched vesicles were originally spherical and became elongated when the solidifying lava was still moving. Amygdules and vesicles are indicators of movement in ancient lavas.

Figure 3.34: Amygdaloidal basalt from the Lake Superior region.

When vesicle pockets in extrusive igneous rocks subsequently filed in, many factors determined whether agate, some other type of silica, or another mineral crystallized inside the empty spaces (see Figure 3.35).

Figure 3.35: Factors and conditions that influenced the type of minerals that filled in vesicle pockets.

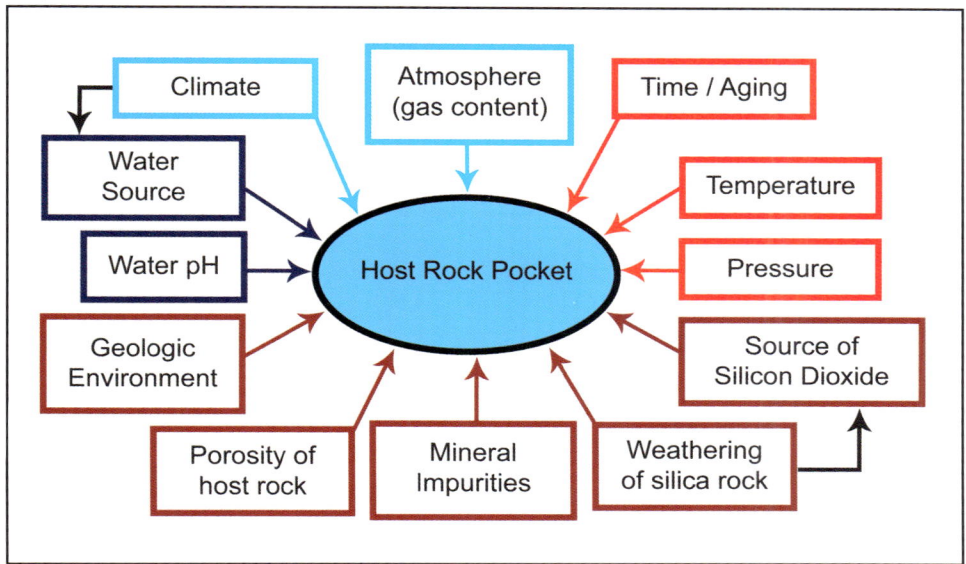

The story of how vesicle pockets filled in is not entirely understood, especially when agates developed. At the time the Midcontinent Rift formed (1.1 billion years ago), the conditions on Earth were different from today. The carbon dioxide content of the Earth's atmosphere was perhaps twenty times higher than the current level. The extreme amount of water vapor released from volcanic activity caused constant rain. Rainwater mixed with carbon dioxide to form carbonic acid (acid rain). Within the rift valley, there were alternating layers of basalt (some of which were vesicular with empty pockets and seams), sediments, volcanic ash, and rhyolite. The acid rain leached through the built-up rock layers and chemically decomposed the materials, freeing up silica and other minerals (see Figure 3.36).

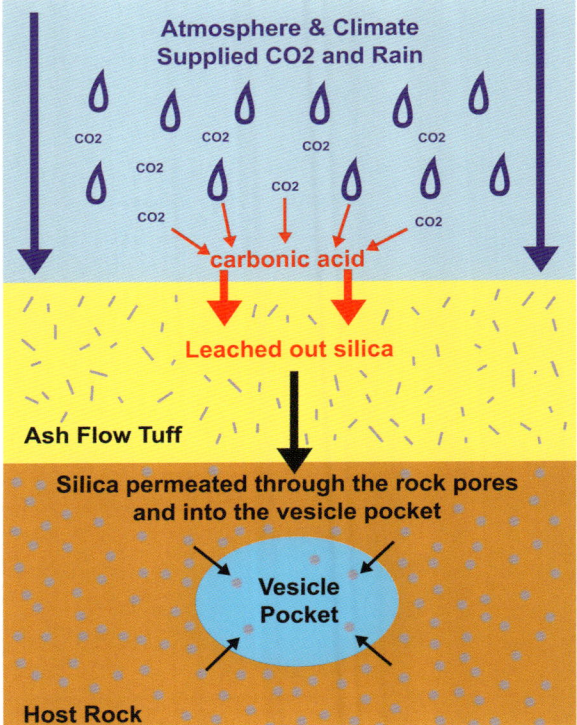

Figure 3.36: This diagram shows how silica was leached from volcanic ash and rhyolite to supply silica that filled vesicle pockets. Not shown is how mineral-rich hydrothermal water also rose from below the Earth's surface to deliver minerals to vesicle pockets.

The ash and rhyolite layers were especially high in silica content. Mineral-rich hydrothermal water also penetrated these layers from below.

Over time, mineral-rich fluids penetrated and filled microscopic pores existing between the crystal grains in basalt, as well as the empty vesicle pockets and fracture seams in the rock. These fluids became supersaturated, causing the formation of small colloid particles. Colloids are homogeneous, non-crystalline particles consisting of large molecules that stay in suspension and do not settle out. These particles at first floated and stayed in suspension, but as the concentration increased, the colloid particles joined. One can think of the polymerizing process as many colloids "holding hands" to form longer and longer chains. This process is an important step in explaining why silica minerals (e.g., chalcedony, agate, etc.) formed in these pockets and seam. It appears that silica does not crystallize directly from single molecules of silicon dioxide suspended in aqueous solutions. Other minerals also developed from these fluids.

The next step in filling the empty pockets was the transportation of colloids into the vesicles. A network of cracks (fractures) that formed in the basalt when it first cooled and hardened served this purpose. In each lava flow, more cooling-induced fractures formed at the top of the flow where the vesicle pockets formed (see Figure 3.37). These fractures served as conduits and facilitated the movement of mineral-rich fluids throughout the host rock (see Figures 3.38 to 3.39). Acid rainwater from precipitation percolated down through the crack system, plus magmatic water from volcanic activity and hydrothermal water from deep within the Earth also transported mineral-rich fluids through the network of cracks. Over time the host rock vesicles filled with mineral-rich water. When the mineral concentration was high enough, minerals crystallized to fill in the vesicle pockets, as well as seams within the host rock.

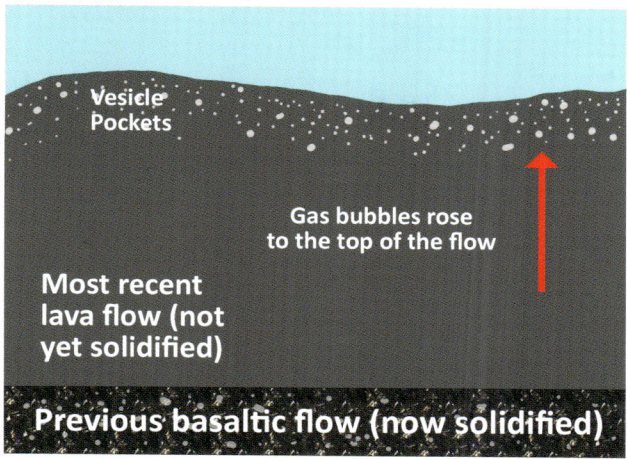

Figure 3.37: Gas bubbles formed vesicle pockets at the top of each lava flow.

Figure 3.38: Mineral-rich water filled vesicle pockets and fractures in host rock.

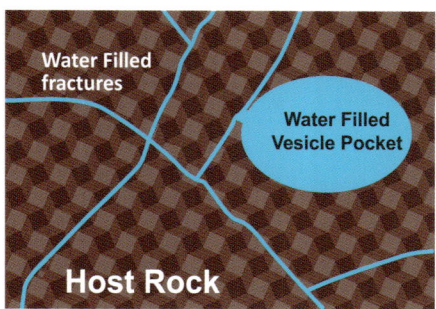

Figure 3.39: This agate-filled vesicle pocket is surrounded by basalt with fractures that fed fluids to the pocket.

Depending on the exact conditions existing when this filling-in process was taking place inside the vesicle pockets, various types of mineral crystals began lining the wall of the vesicle pocket. If the supply of mineral-rich water continued, the entire vesicle pocket filled in. Several types of amygdaloidal minerals often found in extrusive igneous rocks are described at the end of this chapter beginning on page 93.

Amygdaloidal Basalt Photographs

Figure 3.40: Amygdaloidal minerals in basalt including quartz, calcite, and feldspar.

Figure 3.41: The green amygdaloids are filled with epidote.

Figure 3.42: Amygdaloids in weathered basalt.

Rock Identification: Basalt

A specimen can be identified as basalt if it has the following characteristics.

Rock Identification Tips	Photographs or other identification information	
Basalt specimens may have open vesicle pockets, or these pockets may be filled in with amygdaloidal minerals.	*Figure 3.43:* Vesicular basalt cliff with unfilled open cavities.	*Figure 3.44:* Amygdaloidal basalt cliff with mineral-filled cavities.
Some basaltic specimens are porphyritic containing two different sizes of mineral crystals. Larger phenocryst crystals are embedded in a groundmass made of fine-grained mineral crystals.	*Figure 3.45:* Porphyritic basalt from a Lake Superior beach exhibiting two sizes of mineral crystals.	

Rock Identification Tips	Photographs or other identification information
Other than amygdaloids and phenocrysts, the mineral grains in basalt are too small to be seen with the unaided eye.	*Figure 3.46:* This weathered basalt has a very fine-grained texture. 
Basalt is usually black or gray but can be brown if it has a weathered surface.	*Figure 3.47:* Dark gray basalt with amygdaloid minerals.  *Figure 3.48:* This weathered brown basalt specimen has mineral filled amygdaloids. 
Basalt has a harness of 6.0 on the Mohs Hardness Scale and can scratch glass.	*Figure 3.49:* Basalt specimens will scratch glass. 
Differentiate basalt from sedimentary graywacke by the texture. Graywacke (see chapter 4) has angular rock grains cemented together. Basalt has tiny mineral grains molecularly bonded.	*Figure 3.50:* Microscopic image of basalt with two sizes of mineral grains.  *Figure 3.51:* Microscopic graywacke image with different sizes of angular grains. 
Basalt is opaque.	Use a bright flashlight to verify that light does not shine through the specimen.
Basalt rarely contains fossils.	Because basalt crystallized from lava, biologic materials did not usually survive to fossilize.

Basalt Photographs

Figure 3.52: Amygdaloidal basalt with green epidote.

Figure 3.53: Amygdaloidal basalt.

Figure 3.54: Basalt specimen found on a Lake Superior beach.

Figure 3.55: Vesicular basalt.

Figure 3.56: Basalt pebbles on a beach.

Figure 3.57: Amygdaloidal basalt.

Figure 3.58: Basalt with a secondary fill in a fracture seam.

Figure 3.59: Basalt with jasper secondary fills.

Figure 3.60: Porphyritic basalt with phenocryst crystals.

Figure 3.61: Basalt with large amygdaloidal pockets filled with quartz, jasper, and other minerals.

Figure 3.62: Amygdaloidal basalt with agate pockets.

Andesite

Andesite is an extrusive igneous rock that is intermediate between basalt and rhyolite. Andesite is usually light to dark gray in color, due to its content of dark pyroxene or hornblende minerals but can exhibit a wide range of color. Darker andesite is difficult to distinguish from basalt, but even dark colored andesite is usually a bit lighter in color than basalt. When exposed to the elements, andesite weathers to various shades of brown. Due to the tendency for andesite to weather and change its outer color, specimens must be broken for proper examination. Large volumes of andesite, the most common volcanic rock after basalt, formed in island arcs, at continental margin subduction zones, and less commonly, in continental rifts or hot spots, such as the Midcontinent Rift. Even more rarely, andesite is found at mid-ocean ridges. Andesite is rich in plagioclase feldspar (more than 65 percent), silica, and may contain biotite mica, pyroxene, or hornblende. Olivine is rare. It varies from being fine-grained to porphyritic in texture when specimens contain large phenocryst crystals in a fine-grained groundmass matrix. The phenocrysts are usually plagioclase feldspar but can also be pyroxene or hornblende. These high-crystallization-temperature minerals began forming below the Earth's surface and grew to visible sizes before the magma erupted. When the magma erupted onto the Earth's surface, the rest of the lava melt quickly crystallized resulting in a finer-grained groundmass. Porphyritic andesite is the name used for these rocks with two crystal sizes (see Figure 3.63). Andesite is the extrusive cousin to intrusive diorite. Andesite formed from lava with a temperature range from about 1,472°F to 1,832°F (800°C to 1000°C).

Figure 3.63: *Porphyritic andesite with large black hornblende phenocryst crystals in a fine-grained groundmass matrix.*

Andesite formed in many ways. Sometimes within a chamber of semi-molten rock, the heavier crystals separated out and sank to the bottom of the melt. Once these heavier minerals were removed, the remaining melt no longer was basaltic in composition — its silica concentration increased relative to the starting composition. As this process continued, the melt eventually cooled and developed into andesite. In other cases, there was partial melting of the crust that allowed new mixtures of minerals to crystallize and solidify, resulting in the formation of andesite. Andesite most often formed in lava flows produced during active volcanic eruptions. Because these lavas cooled rapidly at the Earth's surface, there was only time for small crystal grains to form. Other than the phenocrysts, the groundmass mineral grains in andesite are usually so small they cannot be seen without the use of magnification. Some specimens that cooled rapidly contain a significant amount of volcanic glass, while others that crystallized from gas-rich lavas may have a vesicular or amygdaloidal texture.

Andesite Photographs

Figure 3.64: *Microscopic image of porphyritic andesite with feldspar phenocrysts.*

Figure 3.65: *Andesite from Nevada.*

Figure 3.66: *Weathered porphyritic andesite from Nevada.*

Figure 3.67: *Weathered andesite.*

Figure 3.68: Weathered andesite that was carved and used as a fishing sinker.

Figure 3.69: Weathered andesite from Slovakia.

Figure 3.70: Gray andesite from Wyoming.

Rhyolite

Rhyolite is a common extrusive igneous rock. It is fine-grained and is either tan, pink, or gray in color (Figure 3.71). Rhyolite often has many unfilled vesicle pockets, and large phenocryst crystals embedded in a fine-grained groundmass matrix. It has a hardness of between 6.0 and 6.5 on the Mohs hardness scale. Rhyolite, like basalt, formed during volcanic events.

Figure 3.71: Porphyritic rhyolite specimen found on a Lake Superior beach.

Most rhyolite developed from granitic lava that had a high silica content. This lava is thicker (more viscous), so it caused explosive pyroclastic eruptions as compared to calmer, effusive lava flow eruptions. Rhyolitic eruptions are more rare than basaltic eruptions. Since 1900 only four rhyolitic eruptions are known to have occurred on Earth. These included the Novarupta Volcano in Alaska (1912), the St. Andrew Strait Volcano in Papua New Guinea (1953 to 1957), and the Chaiten (2008) and Cordon-Caulle Volcanoes in Chile (2011 to 2012).

Volcanic eruptions of rhyolitic lava were more explosive for two reasons. First, the lava was thicker and more viscous because of its high silica content. The difference between the viscosity (thickness) of rhyolitic lavas and basaltic lavas was extreme. Rhyolitic lava at an average temperature of 1472°F (800°C) was ten million times more viscous than room-temperature water whereas basaltic lava at 2012°F (1100°C) was only 100,000 times more viscous. Rhyolitic lava also contained more gases that could only escape through explosive eruptions by blasting the magma from the volcano's vent. Rhyolitic lava was so thick it flowed like toothpaste out of a tube.

Figure 3.72: Rhyolite with phenocryst crystals in a fine-grained groundmass matrix.

Because the rhyolitic lava originated as intrusive magma, some crystals formed slowly in the magma below the Earth's surface and had time to grow large. When the magma rose toward the surface, these crystals stayed intact in the rhyolitic lava (see Figure 3.72). Thus, the components making up rhyolite had two periods of formation: some cooled below the Earth's surface forming phenocryst; others cooled quickly at the surface forming rhyolite's small-grained groundmass. The groundmass is made up of up of between 69 and 80 percent silica, whereas the phenocrysts are mostly potassium and plagioclase feldspars. Rhyolite can also contain small amounts of hornblende, biotite mica, and glassy obsidian shards. Because of the complex dynamics of pyroclastic eruptions, some rhyolite developed flow lines or orbicular patterns as it cooled. Obsidian, pumice, and tuff also formed from rhyolitic lava. The intrusive igneous rock equivalent to rhyolite is granite. Although not as popular as chert and flint, prehistoric people began using a rhyolitic rock from what is now Eastern Pennsylvania over 11,000 years ago. The rock was used to make arrowheads and spear points for hunting because it could be knapped to a sharp point, but it was not used to make weapons — its composition was variable causing it to readily fracture. Although most of the gas pockets in rhyolite remained open and unfilled, mineral-rich fluids sometimes circulated through the rhyolite rock and deposited minerals into the vesicles including opal, jasper, agate, topaz, and red beryl.

Rock Identification: Rhyolite

Rhyolite rock can be found worldwide. Because the mineral components of rhyolite can vary, so can the appearance of rhyolite specimens. Rhyolitic rocks range from a soft tuff to hard obsidian. Because volcanic rocks are classified on the basis of their mineral composition, the fine-grained nature of rhyolite and its diverse appearances make this impractical without significant field experience or sophisticated laboratory equipment. The following characteristics will be true for most but not all rhyolites.

Rock Identification Tips	Photographs or other identification information
When rhyolite solidified from lava at the Earth's surface, the crystallization of the groundmass during cooling cemented all the minerals together.	*Figure 3.73:* The fine-grained groundmass in this microscopic image of rhyolite consists of fused-together minerals. The width of this image is .12 inches (3 mm).

Rock Identification Tips	Photographs or other identification information
Rhyolite is light in color and can be tan, pink, or gray.	*Figure 3.74:* A gray rhyolite. 
Some rhyolite specimens have only one grain size, with the grains being so small they cannot be seen without magnification.	*Figure 3.75:* A monotone-colored pink rhyolite. 
Rhyolite has a hardness of between 6.0 and 6.5 on the Mohs Scale and will scratch glass.	*Figure 3.76:* Rhyolite will scratch glass. 
Some rhyolite specimens have two distinct grain sizes: fine-grained groundmass grains and larger phenocryst crystals.	*Figure 3.77:* This rhyolite specimen has a fine-grained groundmass as well as large phenocrysts. 

Rock Identification Tips	Photographs or other identification information
Because rhyolite forms from slowly flowing lava at the Earth's surface, sometimes the flow pattern is captured in the texture.	**Figure 3.78:** *Lava flow lines are evident in this banded rhyolite specimen.* 
Sometimes orbicular patterns form in rhyolite.	**Figure 3.79:** *Orbicular rhyolite from the Mojave Desert, southern California.* 
Trapped gases in the material from which rhyolite formed often created open vesicle pockets.	**Figure 3.80:** *Vesicular basalt with open pockets.* 
Rhyolite is opaque.	*Use a bright flashlight to verify that light does not shine through the specimen.*
Rhyolite never contains fossils.	*Because rhyolite formed below the Earth's surface from semi-molten magma, it does not contain fossils.*

Rhyolite Photographs

Figure 3.81: A monotone colored rhyolite with phenocrysts and open vesicles.

Figure 3.82: A rhyolite outcrop with complex flow banding.

Figure 3.83: Another rhyolite cliff with flow banding structure.

Figure 3.84: Rhyolite specimen with a large garnet crystal.

Figure 3.85: Rhyolite specimen on a Lake Superior beach.

Figure 3.86: Weathered rhyolite specimen.

Figure 3.87: Pink rhyolite specimens.

Figure 3.88: Porphyritic specimen that formed 640,000 years ago (Yellowstone area, Wyoming).

Figure 3.89: Porphyritic specimen from a Lake Superior beach.

Figure 3.90: Rhyolite specimen with flow banding.

Figure 3.91: Orbicular rhyolite from a Lake Superior beach.

Figure 3.92: Light-colored rhyolite from Colorado.

Figure 3.93: Close-up of a porphyritic rhyolite.

Figure 3.94: A dark-colored specimen of rhyolite.

Figure 3.95: Rhyolite with large feldspar phenocrysts.

Figure 3.96: Like its basaltic "cousin," rhyolite can also form colums.

Figure 3.97: Another rhyolite specimen.

Figure 3.98: Rhyolite with large phenocrysts.

Figure 3.99: Close-up of a rough rhyolite.

Figure 3.100: Collectable wonderstone rhyolite from Utah.

Pumice

Pumice is an extrusive igneous rock that is usually light to medium gray in color and formed from rhyolitic lava. It has a rough vesicular texture, which may or may not contain visible mineral grain crystals. This rock formed when super-heated, highly pressurized rock was violently ejected from a volcano. The "foamy" look of pumice happened because of two simultaneous occurrences. When the material was ejected from the volcano, it underwent both rapid cooling and fast depressurization. The simultaneous cooling and depressurization froze gas bubbles in a glassy matrix. Each vesicle pocket is usually surrounded by a thin wall of volcanic glass. Some specimens show flow lines or banding. Pumice looks similar to a sponge and has a porosity of 64 to 85 percent by volume due to the open vesicles. This extreme porosity causes this rock to float on water, possibly for years, until it eventually becomes waterlogged and sinks. Sometimes large "rafts" of pumice form in underwater volcanic eruptions. For example, several times in the recent geologic past the submerged volcanoes near Tsonga in the South Pacific produced large pumice rafts up to 19 miles in diameter (30 km). Pumice is not commonly found in the Lake Superior region but can be found on the shorelines of other Great Lakes. Pumice is mined in Nevada, Oregon, Idaho, Arizona, California, New Mexico, and Kansas.

Pumice Photographs

Figure 3.101: Light-colored pumice.

Figure 3.102: Pumice with relatively large vesicles.

Figure 3.103: Pumice specimens.

Amygdaloidal Minerals

As previously explained under amygdaloidal basalt, vesicle pockets often formed in extrusive igneous rocks, which sometimes filled in later with minerals. Because these are minerals and not rocks, they will not be featured in this book but will be described. The only mineral for which **Rock Identification** tips are included is agate, which is listed under chalcedony on page 97. This exception has been made due to the high demand for agate identification information. Photographs of all the minerals in this section are included.

Chalcedony

Chalcedony is a microcrystalline silica mineral composed mainly of silicon dioxide but also has other minerals intergrown within its structure (e.g., moganite). There is much confusion about exactly what chalcedony is — the term is not used consistently by geologists or rockhounds. The broadest definition is that all microcrystalline (or cryptocrystalline) silica varieties are chalcedony. A second option suggests chalcedony includes only those specimens that have fibrous micro-crystals, thereby excluding jasper, chert, and flint which have rounded, granular micro-crystals. This definition considers chalcedony to include varieties such as agate, carnelian, chrysoprase, and onyx. The third and most limiting classification is that chalcedony only refers to those specimens with fibrous micro-crystals that have a homogeneous light color (see Figure 3.104). This definition excludes jasper, chert, flint, agate, carnelian, chrysoprase, and onyx. Since geologists use the second definition, it is this option the book has employed.

Figure 3.104: Typical chalcedony with a dense, gray appearance.

In addition to forming in igneous vesicle pockets, chalcedony also developed in silica-rich marine sedimentary rocks, as nodular concretions in limestone rock, as a replacement mineral in other rocks (pseudomorph), and as fossilizing material in petrified wood and silicified fossils. Chalcedony tends to be white, gray, or light blue, is usually translucent, and has a waxy appearance. Some chalcedony specimens contain a small amount of opal, which gives them a watery appearance. Like other silica specimens, chalcedony has a hardness of between 6.5 and 7.0 on the Mohs Scale. It is a bit softer than macrocrystalline quartz and is denser than opal. It fractures in a conchoidal manner like its silica cousins (see Figure 3.105). Chalcedony sometimes fluoresces green under UV light (see Figure 3.106).

Figure 3.105: Diagram of conchoidal fracture. Left image shows top of specimen. Right image shows a side view.

Figure 3.106: Two fluorescent chalcedony rocks from South Dakota, showing top and side views.

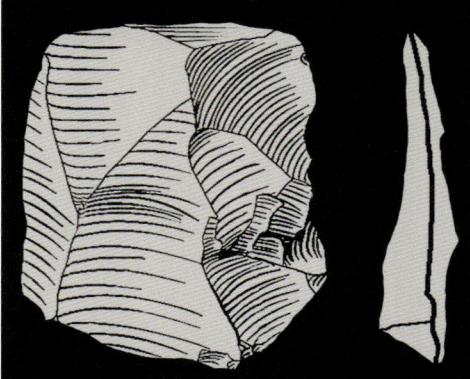

Chalcedony has a wide array of appearances. The fibrous micro-crystals in chalcedony sometimes grew radially resulting in rounded botryoidal formations (see Figure 3.107). It also developed in a stalactic habit (see Figure 3.108) and formed flat discs that expanded out to resemble roses (see Figure 3.109). Sometimes the surface of chalcedony specimens are covered with tiny sparkling macrocrystalline quartz crystals, called drusy quartz, resulting in stones with two different varieties of silica minerals (see Figure 3.110).

Figure 3.107: Botryoidal chalcedony.

Figure 3.108: Stalactic chalcedony.

Figure 3.109: Chalcedony rose.

Figure 3.110: This geode has chalcedony (gray/white) and drusy crystals (red).

Chalcedony Photographs

Figure 3.111: Chalcedony specimen.

Figure 3.112: Blue chalcedony from the Mojave Desert in California.

Figure 3.113: Gray chalcedony.

Figure 3.114: Chalcedony specimen.

Figure 3.115: Blue-gray chalcedony from South Dakota.

Figure 3.116: Another typical gray chalcedony.

Figure 3.117: Complex chalcedony rose.

Figure 3.118: Another chalcedony.

Figure 3.119: Chalcedony with round, botryoidal formations.

Figure 3.120: Another botryoidal chalcedony. This specimen is from Indonesia.

Figure 3.121: Fluorscent chalcedony from South Dakota (365 nm UV).

Figure 3.122: Sometimes specimens are phosphorescent (continue to glow after UV light is turned off).

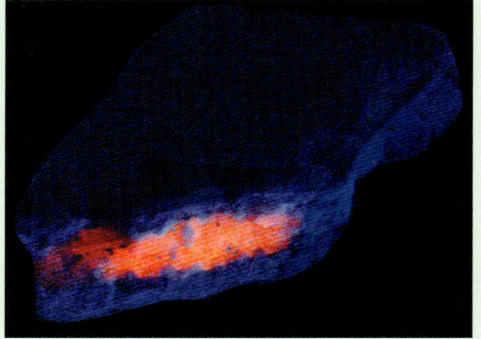

Figure 3.123: Another chalcedony specimen.

Figure 3.124: Chalcedony is sometimes used for carving art.

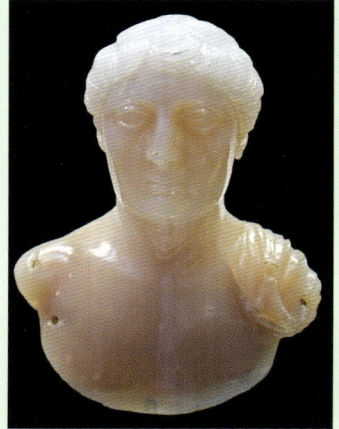

Agate

Agate is a variety of chalcedony that has self-organized banding or other structure. It is estimated there are around 3,000 agate varieties on Earth. Most agates formed inside igneous host rock vesicle pockets and seams (usually in basalt or rhyolite), however, some varieties developed in sedimentary rock. Almost thirty different agate structures have been identified (e.g., concentric, water-level, eyes, sagenite, moss, dendritic, etc.). While scientists have studied agate genesis since the mid-1700s, the agate formation process is still not completely understood. Perhaps this is because no one has documented agate formation in real time, nor have agates been successfully replicated in the laboratory. It is amazing that we cannot make agate, but we can manufacture artificial diamonds, rubies, sapphires, and emeralds.

Although scientists cannot agree on exactly how agates formed, they have concluded that the crystallizing process was dictated by certain self-organizing processes. In other words, the structures within agates resulted from internal dynamics involving the vesicle pocket filling-in process itself, rather than from changing conditions outside the vesicle pocket. Whatever these internal control mechanisms were, they operated worldwide to form a wide variety of agates. The conditions surrounding vesicle pockets may not have controlled agate formation, but they were influential — especially regarding the delivery of silica to the vesicle pockets or marine sediments.

The first step in the filling in of a vesicle pocket for most agates was the formation of a coating of silica on the outside wall of the cavity. Many agates have a "first chalcedony layer" — this band is distinctly different from all subsequent layers. The outer chalcedony layer is often transparent if weathering or post-formation mineral staining has not altered it (see Figure 3.125). The outer layer on agates subjected to significant erosional forces may have eroded away, making it easier to see and identify the internal agate structures in these specimens, such as in water-worn Lake Superior agates and Chinese Rain flower agates (see Figures 3.126 and 3.127).

Figure 3.125: *The outer translucent chalcedony band is clearly visible on the outside of this Brazilian agate slab.*

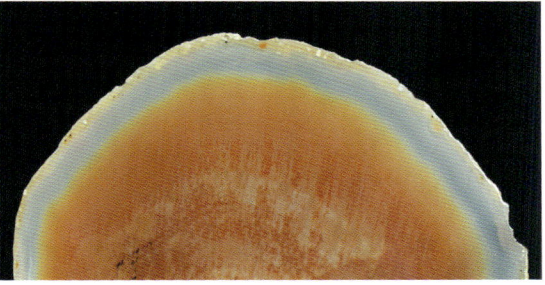

Figure 3.126: Water-worn Lake Superior agate.

Figure 3.127: Water-worn Chinese rain flower agate.

The remaining agate bands inside the first chalcedony layer consist of microscopic, fibrous crystals that grew perpendicular to the orientation of the agate structure. The agate bands developed as fingers of silica grew progressively, layer by layer, to fill the vesicle pocket from the outside wall of the cavity to the interior. Depending on the exact conditions within the vesicle pocket, various types of structures, other than concentric banding, may have also developed. If the silica supply to the pocket was not sufficient, the vesicle pocket did not completely fill in and developed into a geode with a hollow center.

Because of their hardness and durability, agates that have eroded from the matrix rock in which they formed can be found loose as specimens waiting to be discovered by rockhounds on beaches or in gravel pits, rivers, farm fields, and construction sites. In these cases, agates were often transported by water or glaciers great distances from where they formed. In other cases, agates are still sitting right where they originated, such as in Brazil. Agate mining there began in 1830 and continues today. The agates from Brazil are mostly mined from decomposed volcanic ash and basalt that initially was deposited between 250 and 300 million years ago. The "agate mines" are plowed fields or trenches in which loose agate nodules are found and collected.

Mineral Identification: Agate

Although this book features rocks and not minerals, details about how to search for agate mineral specimens will be included due to the high demand for this information. As an example, both the list of agate hunting tips and photos included below will help people find the elusive Lake Superior agate, but these tips can be used to find agates in other geographic locations. Agate hunting is specific to the area and the types of agates found there. Each geographic area is different and requires local knowledge and specifics. To learn about finding agates in your location, visit local rock shops or attend rock club meetings in that region to information seek about where to look and what characteristics are specific to the local agates. Wherever you go agate hunting, it is important to not look at every rock.

When you do this, your brain becomes overwhelmed with all the color and detail, which renders you unintentionally blind. Instead, scan your search area looking for agate characteristics. Let the distinctive agate characteristics find you. Please examine the photos below to see these identification clues for Lake Superior agates.

Rock Identification Tips	Photographs or other identification information
Sometimes agate nodules and seams are still embedded in basalt matrix host rock and have not weathered out.	**Figure 3.128 to 3.130:** *Agate filled in fractures in these basalt specimens. The seam in Figure 3.130 ran out of silica and did not finish filling in.*   
Look for rocks that have red color. The outside surfaces of most Lake Superior agates have some iron oxide staining. This specimen also has conchoidal fractures, gold limonite staining (center left), and possible entrance channels on both ends.	**Figure 3.131:** *Notice the red iron oxide-stained husk of this Lake Superior agate.* 

Rock Identification Tips	Photographs or other identification information
If you find an agate candidate, use either the sun or a bright flashlight to backlight the specimen. Check for translucency as well as for any banding that may not be visible without backlighting. Most rocks are opaque, so translucency is an extremely important characteristic. Sometimes, due to post-formation iron staining, Lake Superior agates only show translucency along the surface edges.	**Figure 3.132:** *Use a bright light or the Sun to check for translucency.* 
Because Lake Superior agates are 1.1 billion years old, the outer husks of most specimens have worn away. For those with husks that have not eroded away, look for evidence of internal structure on the specimen's exterior.	**Figure 3.133:** *This specimen shows possible internal structure (bottom center), iron-oxide red color, and a pit-marked surface.* 

Rock Identification Tips	Photographs or other identification information
Search for rocks with conchoidal fractures that give the specimen an angular, irregular shape. For a diagram regarding conchoidal fractures, please see Figure 3.105 on page 94.	*Figure 3.134:* This raw agate was naturally fractured. 
Look for a pit-marked husk with hollow depressions left from softer minerals that originally lined the outside of the vesicle pocket before subsequently eroding away. Also, notice the mustard-yellow limonite staining in this specimen.	*Figure 3.135:* The pit-marked surface in this agate displays the molded impressions left behind by the softer minerals that eroded away. 
Some Lake Superior agates are not iron-oxide red in color but instead exhibit a dense, gray chalcedony color. This specimen has a distinctive inflow channel (lower right) that fed silica-rich fluids into the vesicle pocket.	*Figure 3.136:* This naturally fractured agate displays the inside of the nodule with dense gray chalcedony color as well as an entrance channel (lower right). 

Rock Identification Tips	Photographs or other identification information
Look for mustard-colored limonite staining. Limonite is a hydrated form of iron oxide. Its color can vary from yellow to orange.	**Figure 3.137:** This agate has mustard-yellow limonite staining over gray chalcedony. 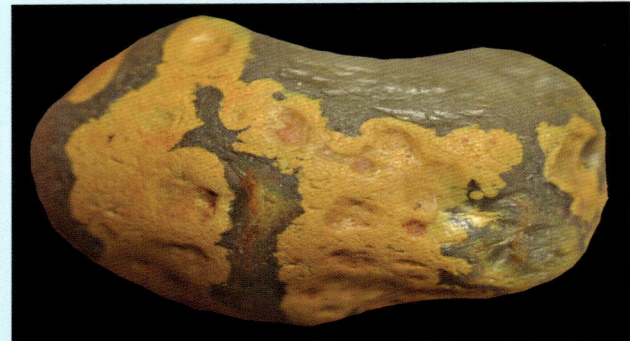
Look for rocks that show evident concentric banding or other structure.	**Figure 3.138 to 3.139:** These raw agates show structure on the outside.  
Scan for specimens that have a dense, waxy luster. The top two specimens are peelers that exhibit differential erosion across the agate bands. All three have iron-oxide staining. The bottom right specimen has a pit-marked surface.	**Figure 3.140 to 3.142:** These specimens show a dense, waxy luster.   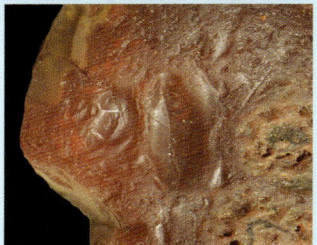

Rock Identification Tips	Photographs or other identification information
There are more than 30 types of agate structure. The most common is fortification/ concentric banding. There is also grape, brecciated, candy-striped, crazy lace, dendritic, enhydro, eye, floater band, iris, moss, paintstone, sagenite, shadow, tube, water-level and more. Not all structure types are covered in this book.	*Figure 3.143 to 3.145:* These specimens show horizontal water-level banding, eye formations, and sagenite structure. 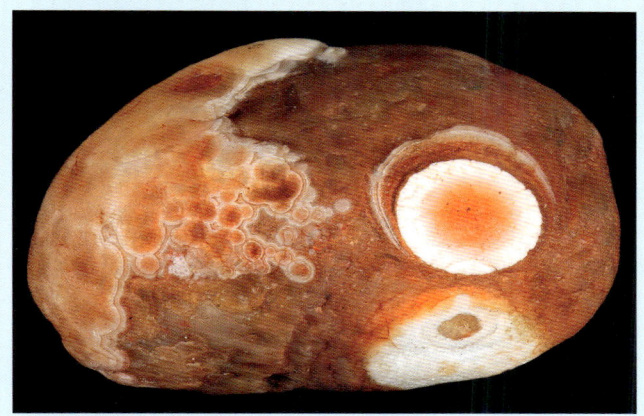

Carnelian

Carnelian is a variety of chalcedony with color ranging from light yellow to dark red. Carnelian also has microcrystalline silica with crystals so small you need a scanning electron microscope to see them. As a result, these microcrystalline silica specimens have a very dense appearance. The color in carnelian occurs because as the microcrystals were forming, particles of iron-oxide became trapped throughout the structure. Many carnelian specimens have the same color throughout. However, carnelian can develop with banding or other structure and is then classified as a carnelian agate.

Figure 3.146: *This Chinese urn was carved from one piece of carnelian.*

Carnelian formed in pockets and seams of igneous rocks as well as in sedimentary and metamorphic rock deposits. Because carnelian is harder than the surrounding rock in which it formed, over time various geologic, weathering, and chemical forces eroded the host rock freeing and releasing carnelian nodules. Carnelian stones were then transported away from the source by streams and glaciers and were deposited in riverbeds, glacial till, farm fields, and beaches.

Because of its beautiful color and translucency, carnelian is classified as a semi-precious gemstone. The darker the specimen, the less translucency there is because the darker color is from iron oxide in the microcrystalline structure. Carnelian is quite hard (7.0 on the Mohs Scale) and can easily be fashioned into jewelry and other items (see Figure 3.146). Carnelian is the national stone of both Norway and Sweden. It is also the traditional gemstone gift for the 17th wedding anniversary. Humans have been collecting and using carnelian since the dawn of civilization. Historically, common people could not afford jewelry made with carnelian or other semiprecious minerals, so this stone was a status symbol for the wealthy and powerful. In addition to making jewelry, carnelian was used to imprint seals on wax to secure envelopes and legal documents because carnelian does not stick to wax.

An interesting historical fact about carnelian involves Napoleon Bonaparte. While he was in Egypt during his travels, he acquired a carnelian amulet. He believed it to be good luck and carried it with him for the rest of his life. Napoleon passed the amulet on to his nephew, Napoleon III, who also wore it for the rest of his life. He, too, believed in its "magic" and survived numerous assassination attempts. It was then passed to his son, but this Napoleon did not believe in such superstitions and only wore it sporadically. When he was killed by Zulus in Africa, the amulet was not on his body and has not been found since.

Carnelian Photographs

Figure 3.147: Carnelian "chippers," some of which are carnelian agates.

Figure 3.148: A polished carnelian stone.

Figure 3.149: A polished carnelian agate.

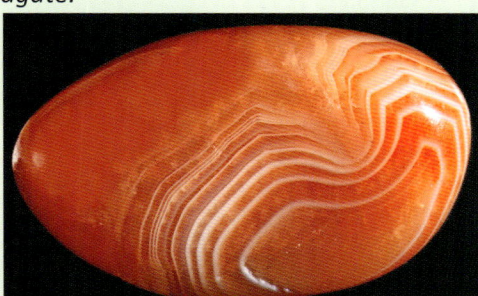

Figure 3.150: Carnelian is almost always translucent.

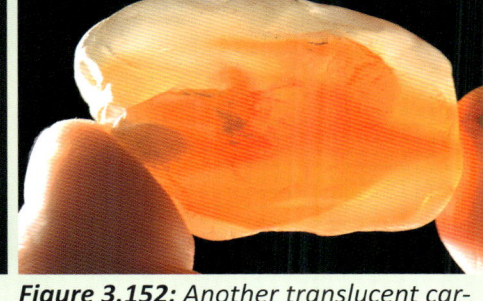

Figure 3.151: Some of the fracture seams in this basalt contain carnelian.

Figure 3.152: Another translucent carnelian.

Figure 3.153: The top of this seam agate is dark-red carnelian.

Figure 3.154: This two pound (1 kg) carnelian agate was found in Grand Marais, Michigan.

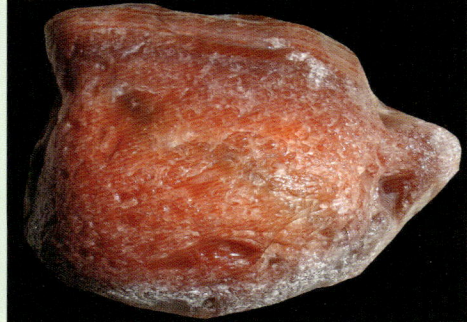

Figure 3.155: This agate has some carnelian sections.

Figure 3.156: A group of carnelian stones.

Figure 3.157: Some rough and polished carnelian stones.

Figure 3.158: This carnelian brooch was made for a Ptolemaic queen in 200 B.C.

Calcite

Calcite specimens that started as amygdaloidal infills in basalt rarely survived as collectable specimens when they eroded free of the matrix rock due to this mineral's softness. Calcite amygdaloids usually are fairly translucent, can be tabular or rhombic in shape, and are colorless to white in color. They can exhibit a wide variety of colors if the amygdaloid included minor impurities. Vein deposits of calcite are more common in the basalts and gabbros along the North Shore of Lake Superior (see Figure 3.159). Calcite is soft with a hardness of 3.0 on the Mohs Scale, so it can be scratched with a metal nail. This differentiates calcite from silica amygdaloids such as chalcedony or quartz, which cannot be scratched with a nail. Calcite inclusions will react and bubble or fizz with vinegar because the acid causes a chemical reaction that gives off carbon dioxide.

Calcite Photographs

Figure 3.159: Calcite crystals formed in this Minnesota basalt outcrop seam.

Figure 3.160: The white amygdaloids in this basalt are calcite; the green are epidote.

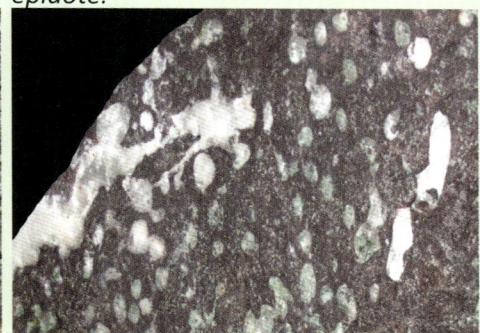

Figure 3.161: Calcite infills dominate this basalt specimen from the Keweenaw Peninsula (Michigan).

Chlorastrolite (Isle Royale Greenstone)

Chlorastrolite, also known as Isle Royale greenstone, is the gem variety of pumpellyite. Specimens are green to bluish-green with a distinctive and attractive mosaic or segmented pattern, sometimes referred to as a "turtleback" pattern (see Figure 3.162). The mineral can exhibit a chatoyant luster, which means that light reflecting off the surface has a shimmering or glimmering variance when the specimen is rotated in either sunlight or a bright light. This chatoyant luster is due to fibrous structures in the rock that scatters light, similar to the changing luster of the eyes in cats. Chlorastrolite occurs as amygdaloid and fracture infills in basalt, and as beach pebbles in loose sediments when eroded from the basalt. This stone is found exclusively in the Keweenaw Peninsula of the Upper Peninsula of Michigan and on Isle Royale in Lake Superior. Isle Royale is a National Park, so it is illegal to collect specimens there. It is difficult to identify an unpolished pebble of chlorastrolite (see Figure 3.163). Most gem quality chlorastrolite stones are very small, usually the size of a pea, and almost always smaller than a half inch (1.27 cm). One of the largest gem quality chlorastrolite is in the Smithsonian Museum and measures 1.5 by 3.0 inches (3.80 by 7.62 cm). Chlorastrolite was designated as the official gemstone of the state of Michigan in 1973. These "greenstones" have a hardness of 5.0 to 6.0 on the Mohs scale and are very well suited for all varieties of jewelry. Polishing is tricky, however, as some stones have softer chlorite or calcite centers or can even be hollow. Greenstones are also very difficult to extract from basalt without breaking them.

Figure 3.162: Chlorastrolite specimen.   *Figure 3.163:* Rough weathered basalt with embedded Isle Royale greenstones.

Chlorastrolite Photographs

Figure 3.164: Rough chlorastrolite polished in "situ" in basaltic host rock.

Figure 3.165: Another chlorastrolite polished in situ.

Figure 3.166: A group of polished chlorastrolites.

Copper

Copper is a chemical element (Cu on the Periodic Table of Elements) with an atomic number of 29. It is malleable and ductile, which means it can be hammered out of shape without breaking or losing its toughness. Copper is one of the few metals that can naturally occur in pure form that can directly be used as a metal. This property led to early use by humans of copper beginning around 8,000 B.C. Copper also has high thermal and electrical conductivity, so it is used in making thermocouples for temperature measurement as well as in electrical wiring components. It is also mixed with other elements to make metal alloys such as sterling silver and bronze.

Copper is one of a few metallic elements with a natural color other than gray or silver. Pure copper is orange-red in color (see Figure 3.167) and acquires a reddish or green tarnish when exposed to air. Copper does not react with water, but it does slowly react with atmospheric oxygen to form a layer of brown-black copper oxide that protects the underlying metal from further corrosion, unlike rust that forms on both surface and underlying layers of iron (see Figure 3.168). A green layer of copper carbonate is often seen on old copper structures, such as on the Statue of Liberty (see Figure 3.169).

Figure 3.167: Copper specimen with the outer oxidized coating cleaned off.

Figure 3.168: Oxidized copper specimen.

Figure 3.169: Copper carbonate coats the Statue of Liberty.

Indigenous people mined and used copper for millennia before explorers reached the Upper Peninsula of Michigan. The first copper mine operated by Europeans was opened in 1771, but speculative activity did not begin in earnest until 1841. Between 1843 and 1846, thousands of prospectors arrived in Michigan's Upper Peninsula with the hope of "striking it rich." These mines operated commercially for more than a hundred years producing over 14 billion pounds (6.3 billion kg) of copper, which helped fuel the industrial revolution. Some of the copper deposits were remarkable for their purity, quality, and size measuring between two and eight miles wide (3.2 to 12.9 km)! Unfortunately, when underground mining became cost prohibitive, all Michigan mining of native copper stopped in 1969. Currently, Arizona leads the nation in copper production via the use of large open-pit mines.

Copper was deposited in Michigan rocks when hot hydrothermal waters percolated upward from deep below the Earth's surface carrying copper and other minerals in solution. These mineral-rich waters filled the vesicle cavities and fracture seams in both volcanic and sedimentary rock layers with pure copper within the Midcontinent Rift, making it the greatest deposit of native copper in the world. Copper was also deposited in cracks within the Copper Harbor Conglomerate layers (see Chapter 4).

Most all the reachable copper has been mined in the Upper Peninsula. However, there are still some basalt rocks that can be found with copper amygdaloids. Old mine dump material has been crushed to make road gravel, so metal detectors can be used to search roadsides in the Keweenaw Peninsula for pieces of copper. Float copper, which are glacially transported specimens, can also be found in the western Upper Peninsula.

Copper Photographs

Figure 3.170: Amygdaloidal basalt with copper amygdaloids.

Figure 3.171: Cut slab of amygdaloidal basalt showing copper amygdaloids.

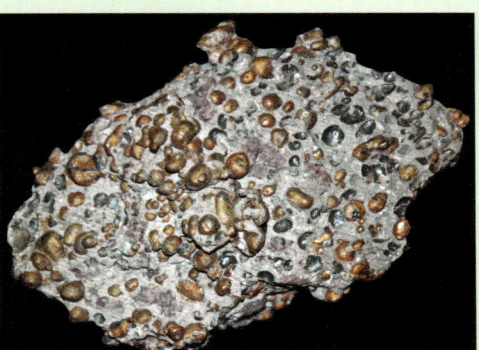

Figure 3.172: This cut slab of conglomerate has copper that filled in as an intergranular cement.

Figure 3.173: Cupriferous amygdaloidal basalt (copper ore).

Figure 3.174: Glacial float copper boulder.

Datolite

Datolite is another mineral that filled vesicle pockets in basalt. This was especially true in what is now the Keweenaw Peninsula of Michigan. The datolite deposits were associated with copper-rich hydrothermal fluids, which upwelled through the Midcontinent Rift valley a billion years ago. Unlike most datolite found elsewhere in the world, the Michigan datolite is fine-grained in texture and has attractive coloration due to the inclusion of copper and associated minerals during hydrothermal fluid precipitation.

Figure 3.175: Whole datolite nodule with cauliflower-like husk.

Datolite has a vitreous luster with a shiny appearance similar to glass. It can be found as amygdaloids in basalt, or as round cauliflower-looking nodules that have eroded free of the matrix basaltic rock (see Figure 3.175). Most nodules are thick husked and do not reveal the datolite on the inside — they usually must be cut in half to expose the mineral. Datolite can be white, gray, pale green, light red, yellow, or pink. It has a hardness of 5.0 to 5.5 on the Mohs Scale. Datolite can be translucent or opaque.

Datolite Photographs

Figure 3.176: Keweenaw datolite, cut and polished (Michigan).

Figure 3.177: Datolite with copper.

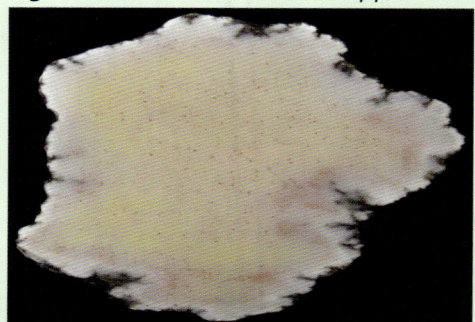

Figure 3.178: Another cut Keweenaw datolite (Michigan).

Epidote

Epidote is a pistachio-green colored mineral found in basalt amygdaloidal vesicles as well as in nearly every type of metamorphic rock. It can fill in cracks and seams, grow as crystals, or develop into thin green crusts (see Figure 3.179). This mineral formed when igneous rocks contacted hot hydrothermal water during metamorphism, which caused some of the mineral grains in the igneous rock to change and be replaced with epidote. Although it almost always is green in color (ranging from yellowish green to dark green), it can also be brown and black. The epidote that formed in igneous rock amygdaloidal pockets are sometimes striated (grooved). It has a hardness of 6.0 to 7.0 on the Mohs Hardness Scale, so it will scratch glass. It is usually opaque, but it can have translucent components. Specimens can be almost all epidote, or they can have other minerals mixed in such as feldspar and quartz. Although pure epidote crystals can be translucent, epidote amygdaloids and epidote beach stones are opaque. Epidote is one of the key components of unakite (see Chapter 5).

Figure 3.179: Pistachio-green epidote.

Epidote Photographs

Figure 3.180: Epidote amygdaloids in basalt from the Keweenaw (Michigan).

Figure 3.181: Epidote seam in Virginia host rock.

Figure 3.182: Epidote secondary fill in granite.

Figure 3.183: Epidote specimen from a Lake Superior beach.

Figure 3.184: A group of epidote specimens.

Figure 3.185: Epidote coats this Massachusetts rock.

Figure 3.186: Granite metamorphosed into mostly epidote.

Figure 3.187: This cross-polarized microscopic image of epidote is magnified 10 times (field of view=.08 inches/2 mm).

Figure 3.188: Epidote with some pink potassium feldspar, thus, approaching unakite (see chapter 5).

Milky Quartz

Milky Quartz is a macrocrystalline variety of silica. It is the most common type of crystalline quartz. Quartz can form large, translucent, six-sided crystals, or it can develop within rock cavities into small (but visible) crystals of drusy quartz (see Figures 3.189 to 3.190). Milky quartz is any quartz specimen or crystal cluster that is white in color and cloudy (translucent but not transparent). Macrocrystalline quartz grows by adding molecules to the crystal's surfaces, layer by layer. This is different from the cryptocrystalline quartz varieties (e.g., chalcedony), that forms individual microscopic silica crystals from colloidal fluids, which are liquids with suspended silica particles. The white color is caused by minute inclusions of gas, liquid, or both, that became trapped during crystal formation. These inclusions make milky quartz of little value for optical and lapidary gemstone applications.

Figure 3.189: Clear macrocrystalline quartz crystal.

Figure 3.190: Drusy quartz.

The milky quartz pebbles found in many environments have been rounded and smoothed by erosional forces (see Figure 3.191). When you rotate a milky quartz pebble in bright light, you can usually see the glitter-sized crystal faces. Since milky quartz formed quickly from hydrothermal fluids, there was a temperature differential between the outside and inside of the silica structure. This temperature difference often caused thermal fractures, which are evident throughout many milky quartz specimens (see Figure 3.192). Milky quartz has a hardness of 7.0 on the Mohs Scale, so it easily scratches glass.

Figure 3.191: Milky quartz pebbles found on a beach are often smooth and round by erosional forces.

Although it is chemically resistant, quartz is weakly soluble so most natural waters contain some dissolved silica that can precipitate as quartz. Thus, quartz is a common vein-filling and cavity-filling mineral in rocks through which silica-rich waters have moved. There are numerous rocks that contain these white secondary fills (see Figure 3.193). Secondary fills occurred when minerals filled thermal fractures within the host rock. A "wishing stone" is a rock with a single secondary filled vein that circles entirely around the stone without any breaks. If you are lucky enough to find a wishing stone: pick it up, close your eyes, make a wish, and carefully throw the stone as far as you can (especially if you are next to a lake). Some people believe this practice will make their dreams come true.

Figure 3.192: Milky quartz with visible thermal fractures.

Figure 3.193: Wishing stones have one secondary fill seam circling the rock.

Milky Quartz Photographs

Figure 3.194: Milky quartz seam in basalt.

Figure 3.195: Amygdaloidal basalt with mineral infills including milky quartz.

Figure 3.196: Milky quartz pebble found on a Lake Superior beach.

Figure 3.197: Milky quartz pebble with iron-oxide staining and thermal fractures.

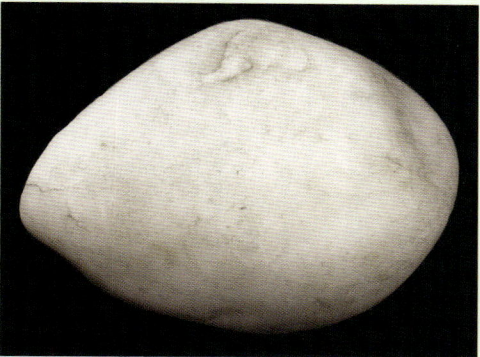

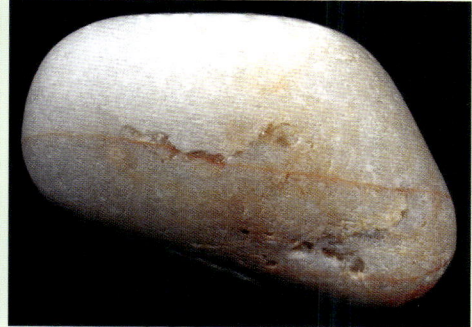

Figure 3.198: Rough specimen of milky quartz (not found on a beach).

Figure 3.199: Gold can be associated with milky quartz.

Figure 3.200: Specimens of milky quartz found on a beach.

Figure 3.201: Rough milky quartz specimen from Arizona.

Prehnite

Prehnite is an abundant secondary mineral found in basaltic matrix rocks on Isle Royale, in the Keweenaw Peninsula, and elsewhere in the Lake Superior region. It occurs as amygdule and fracture fillings, and as a replacement of earlier minerals in rock. Prehnite is often found associated with minerals such as thomsonite, datolite, calcite, and others. This mineral was delivered by hydrothermal fluids to the cavities and veins in basaltic and andesitic rocks. Prehnite often formed as radial crystal aggregates causing it to be misidentified as thomsonite (see below). Most prehnite is pale apple-green, but it can also be nearly black, pink, gray, yellow, and off-white. Prehnite can have intergrowths of copper and silver. Specimens with chlorite and specks of native copper are named "patricianite," which was named by a Michigan resident in 1966 after his daughter. Prehnite has a hardness of between 6.0 and 6.5 on the Mohs Scale so it will easily scratch glass. A fresh surface of prehnite has a vitreous, pearly luster while surfaces have a white husk. Specimens can be opaque to translucent.

Prehnite Photographs

Figure 3.202: Prehnite from the Keweenaw Peninsula in Michigan.

Figure 3.203: This specimen has prehnite mixed with other minerals.

Figure 3.204: Prehnite formations on basalt.

Thomsonite

Thomsonite is a zeolite mineral found along the north shore of Lake Superior and in the Keweenaw Peninsula of Michigan. It is often confused with prehnite, which can have similar colors and radiating patterns. It has a hardness of 5.0 to 5.5 on the Mohs Scale and is transparent to translucent. It can be colorless, white, beige, green, yellow, or red — but usually has a mixture of colors. The crystals tend to be long thin blades that often formed round, radial patterns or fan-shaped arrangements.

Thomsonite occurs with other zeolites in the amygdaloidal cavities of basalt, and occasionally in granitic pegmatites. In addition to the Lake Superior region, thomsonites are also found in Arkansas, Colorado, New Jersey, Oregon, and a few other places around the world. Most thomsonite nodules are found still embedded in basalt or other matrix rock. They are quite small, around .25 inch (.6 cm), and are very brittle. Consequently, they are difficult to extract from the matrix rock without breaking them. Specimens used as gemstones are usually intact pebbles collected from beaches.

Thomsonite Photographs

Figure 3.205: Thomsonite amygdaloids are peaking out of this basalt specimen.

Figure 3.206: A collection of thomsonites.

Figure 3.207: Close-up of thomsonites from Minnesota.

Figure 3.208: Another close-up of Minnesota thomsonites.

Figure 3.209: Close-up of a pink thomsonite.

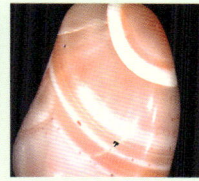

Chapter 4: Sedimentary Rocks

Note: For a list of the Featured Rocks included in this chapter, please go to page 123.

Geologic Background

Sedimentary rocks formed over time through the deposition, compaction, fluid removal, and cementation of sediments. Most sediments are solid loose particles of material that broke off during the weathering of rock. Both chemical and mechanical forces broke large rocks into smaller pieces. All rock weathers over time. Weathering was (and is) distinctly different from erosion, which involved transportation of the loose weathered particles away from their source areas. Erosional transport was facilitated by water, wind, gravity, or ice. The distance moved between source rock and depositional zone may have been a few feet or thousands of miles. These sediments — which ranged in size from boulders to cobbles, gravel, sand, silt, and clay — were transported until there was no longer enough energy to move them. When the sediments reached their deposition zones, they settled out onto the bottoms of rivers, lakes, and oceans. Water is the primary mover of sediments and can transport particles of any size if the currents are strong enough. Wind moves lesser amounts of sediments, only transporting dust-to-sand-sized particles. Glacial ice can move massive amounts of material including car-sized blocks of rock, but this particle transport method has been limited by the few number of ice ages over geologic time and the geographically limited number of glaciers. Today glaciers only cover about three percent of the Earth's surface.

During transport, the power of moving water arranged sediment particles by size, and to a lesser degree by shape. This is called sorting. Sediments transported longer distances tended to be more worn down and better sorted. Sediments not sorted well are called unconsolidated. The four types of sediments are:

> **Clastic sediments** form through the processes of weathering, breaking apart, and erosion of pre-existing rock. Clastic sedimentary rocks are divided into several groups based on the size of the sediments including: clay/silt are less than .002 inches in size (1/16 mm) and are not visible to the unaided eye; sand is between .002 and .08 inches (1/16 and 2 mm); gravel is greater than .08 inches (2 mm).

> **Biologic sediments** form by the accumulation of organic material or by biologic activity. Most biologic sediments are composed of the remains of living organisms such as phytoplankton (microscopic algae), zooplankton (animals), terrestrial and aquatic plants, invertebrate shells, vertebrate remains (teeth, bone), and associated organic residues (e.g., coal and limestone).

Chemical sediments directly precipitate from mineral-rich water. Examples include rocks called evaporites that formed when water evaporated and left behind minerals, which built up over time to form rock (e.g., salt and gypsum).

Cosmogenous sediments originate from outer space. Scientists estimate that between 100 and 300 tons of cosmic material, mostly dust, hit Earth each day. Cosmogenous sediments can be microscopic spherules or larger rock debris from meteors. Both types can be composed of silica, iron, nickel, and other minerals.

Over time, layers of sediment pile up in depositional settings. As sediments accumulate, older sediments become buried under younger sediments. When layers pile up enough, the weight of overlying sediments cause the lower layers to compact (see Figure 4.1). As the weight of overlying rock increases, higher pressures expel fluids from pores between the sediment grains. The pressure and higher temperatures at depth combine to cause chemical and physical changes to the buried sediments (diagenesis). Minerals precipitate from dwindling solutions and serve as a natural glue to bind sediment particles together (lithification). This process turns loose sediments into solid sedimentary rocks. Common cementing minerals include silica, calcite, limonite, hematite, and clay.

Figure 4.1: *This diagram shows the process from deposition to compaction to rock formation.*

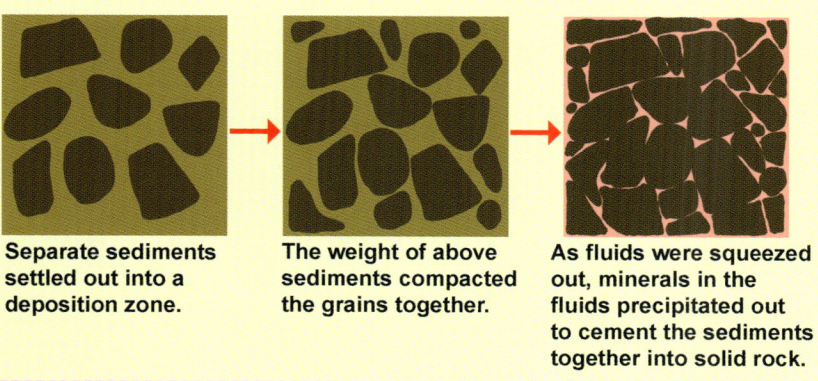

Separate sediments settled out into a deposition zone.

The weight of above sediments compacted the grains together.

As fluids were squeezed out, minerals in the fluids precipitated out to cement the sediments together into solid rock.

Coal is an example of a sedimentary rock that formed when increased pressures and temperatures at depth stimulated additional chemical reactions. In this case, biologic sediments solidified and were chemically transformed into coal (see Figure 4.2). If this continuum of higher temperatures and pressures happened to the extreme, even more drastic metamorphic changes would have happened (see Chapter 5 for information about Metamorphic rocks).

Figure 4.2: Bituminous coal – a type of sedimentary rock.

Sedimentary rocks cover around 73 percent of our planet's current land surface. But because sedimentary rocks develop at or near the surface, this rock type makes up only eight percent of the Earth's crust by volume. Most sedimentary rocks formed in coastal environments along the edges of continents and in shallow seas that encroached and flooded continental basins. While most sedimentary rocks formed in the last few hundred million years, some varieties date to three billion years old. Sedimentary rocks have preserved the fossil record which have helped to determine when, where, and which ancient life forms lived and died (see Figure 4.3). By matching up fossil records on different landmasses, fossiliferous rocks have helped scientists to determine the movement of tectonic plates around our planet. Sedimentary rocks also preserved evidence of ancient environments and changing landscapes.

Figure 4.3: Trilobite fossil.

Unlike igneous rocks, sedimentary rocks contain significantly fewer minerals. Most sedimentary rocks consist of either silica or calcite. Other minerals in the original sediment deposits were not chemically stable, and instead often weathered into clay minerals.

Featured Rocks

The sedimentary rocks featured in this chapter include:

> **Key**
> - Featured rock (includes rock identification information and photographs).
> - Rock is described but not featured (includes photographs).
> * Varieties of these Great Lakes rocks are included as examples.

- **Banded Iron Formation** (page 124): A distinctive sedimentary rock consisting of alternating layers of iron-rich ore and silica-rich chert or jasper;
- **Breccia** (page 130): A clastic sedimentary rock with sharp, angular fragments;
- **Chert** (page 137): A fine-grained, opaque sedimentary rock;
    - Jasper (page 143): A variety of chert;
- **Conglomerate** (page 147): A clastic sedimentary rock with rounded fragments;
    - Copper Harbor Conglomerate* (page 148): An ancient conglomerate rock found in the western Lake Superior region;
    - Puddingstone (page 150): A conglomerate rock with rounded clastic pebbles that contrast in color with the fine-grained matrix;
- **Sandstone** (page 154): A clastic sedimentary rock made of fused sand grains;
    - Graywacke (page 156): An ancient, dark-colored sandstone;
        - Omarolluk (Omar)* (page 159): Graywacke rocks with distinctive, circular indentations;
    - Jacobsville Sandstone* (page 161): This ancient sandstone lies around and under Lake Superior;
    - Munising Formation Sandstone* (page 168): This sandstone makes up most of Michigan's Pictured Rocks National Lakeshore;
- **Limestone** (page 172): A fine-grained to coarse-grained sedimentary rock formed from calcium carbonate;
- **Shale** (page 179): A fine-grained, clastic sedimentary rock composed of fused mud, clay minerals, and silt-sized particles.

Banded Iron Formation

Banded Iron Formations, or BIFs, are unusual, dense sedimentary rocks consisting of alternating layers of iron-rich rock and silica-rich rock. Most BIFs formed between 3.0 and 1.8 billion years ago. One exception is a BIF deposit recently found in China, dating to 527 million years old. This deposit formed from upwelled, iron-rich hydrothermal fluids, which is different than the formation process resulting in older BIF deposits (described below). Many specific varieties of iron formation are known, and some are given special rock names. For example, jaspilite is an attractive reddish and silvery gray banded rock consisting of alternating bands of either hematite or magnetite and red chert/jasper (see Figure 4.4).

Figure 4.4: Jaspilite, a variety of a BIF from Western Australia. Field of view is 4.2 inches wide (10.7 cm).

When BIFs started to form about three billion years ago, oceans were green from a high concentration of dissolved and uncombined iron. The atmosphere was orange from the absolute lack of oxygen and a high concentration of methane. When the first photosynthesizing organisms evolved on Earth around 3.5 billion years ago — cyanobacteria (see Figure 4.5) — everything on our planet began to change. These primitive organisms used sunlight to synthesize food from carbon dioxide and water. A by-product of their photosynthetic activity was oxygen. Because the cyanobacteria lived in shallow oceans, the iron molecules that were dissolved in the oceans immediately started combining with the newly available oxygen to form iron oxide. The heavy iron oxide molecules precipitated out of solution and sank to the seafloor. Blooms in the populations of cyanobacteria produced layers of organic remains when the organisms died, which also sank to the sea floor. Ironically, the early species of cyanobacteria could not survive in an oxygen-rich environment, which is why they had massive die-offs. The layers of Banded Iron Formation developed due to an alternating pattern of cyanobacteria blooms and die-offs producing silica sediments, and iron oxide sediments produced by the combining of free iron with oxygen.

When nearly all the free iron in the oceans was chemically combined, additional oxygen produced by cyanobacteria was able for the first time to accumulate in Earth's atmosphere. The increasing concentration of oxygen in the atmosphere helped facilitate the evolution of more complex life forms and allowed an ozone layer to form. This atmospheric layer protects the planet and life forms from harmful solar radiation. Once the oceans and atmosphere were oxygenated, they became the blue colors seen today.

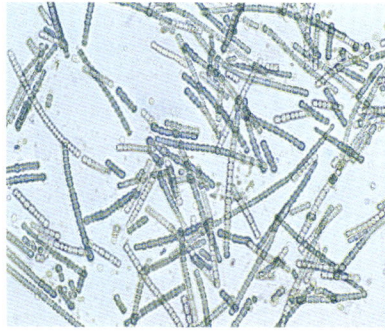

Figure 4.5: Magnified image of modern cyanobacteria.

BIF layers found worldwide record the oxygenation of our planet. Cyanobacteria dominated life on Earth for about a billion years. Over time, species of cyanobacteria evolved to be tolerant of oxygen-rich environments, after which the cyanobacteria species populations exploded. Because of their age, most BIFs have been around long enough to have been subjected to one or more orogenic (mountain-building) events caused by plate tectonic forces. As such, most BIFs are folded and/or metamorphosed to varying degrees. Some of the most famous and extensive BIF deposits are found in the vicinity of North America's Lake Superior Basin (see Figure 4.6). These BIFs have economic concentrations of iron and have been or are still being mined. BIFs are the most important source of iron ore on Earth.

Figure 4.6: Location of Lake Superior iron ranges.

Figure 4.7: The summit of Jasper Knob, located in Ishpeming, Michigan.

One of the most extensive BIF units is the 1.874-billion-year-old Negaunee Iron Formation, which outcrops in the Marquette Iron Range of Michigan's Upper Peninsula (see Figure 4.7). This deposit was discovered in 1844 and has been mined continuously since 1847. It is about 2,500 feet thick (760 m). The Negaunee deposit still has economic concentrations of iron, but much of the unit in the Marquette Iron Range has been mined out. The best remaining, easily accessible outcrop to visit and see is Jasper Knob in the town of Ishpeming, Michigan. Earth's oldest known macrofossils, *Grypania spiralis*, occur in this unit (see Figure 4.12).

Rock Identification: Banded Iron Formation

A specimen can usually be identified as Banded Iron Formation if it has the following characteristics. **Note**: BIF specimens can vary in appearance.

Rock Identification Tips	Photographs or other identification information
BIFs consist of repeated, thin layers of silver to black iron-rich deposits, either magnetite (Fe_3O_4) or hematite (Fe_2O_3), alternating with bands of silica-rich chert or jasper, often red in color.	*Figure 4.8: Banded Iron Formation rocks have alternating layers.* 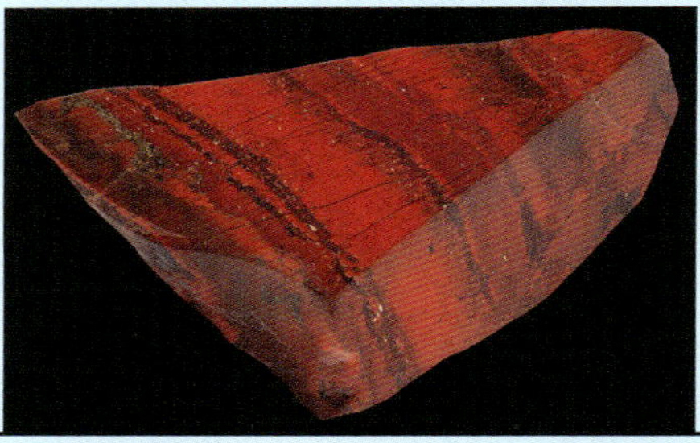
Because BIF deposits are more than 1.5 billion years old, the deposits have been subjected to plate tectonic forces and many have been folded and metamorphosed to some degree.	*Figure 4.09: The folding of the layers in this BIF rock was caused by plate tectonic forces.* 
Although many BIF specimens contain hematite, those that contain magnetite will attract a magnet.	*Figure 4.10: Use a magnet to determine if the BIF specimen has magnetite.* 

The hardness of BIF varies ranging from 5.5 to 6.0 on the Mohs Scale. Usually a specimen will scratch glass, but a nail will always scratch a BIF specimen.	*Figure 4.11:* A nail will scratch Banded Iron Formation specimens. 
The percentage of iron minerals in BIF specimens varies between 30 and 60 percent. Thus, BIF specimens are very heavy for their size.	*Compare the weight of a BIF specimen with that of another type of rock the same size: the BIF specimen will be heavier.*
Most BIF specimens do not contain fossils, but a few may have very primitive fossils. Some stromatolite specimens can also be BIF.	*Figure 4.12:* Grypania spiralis *tube-shaped fossils are evident in this 1.874 billion-year-old BIF from the Upper Peninsula in Michigan.* 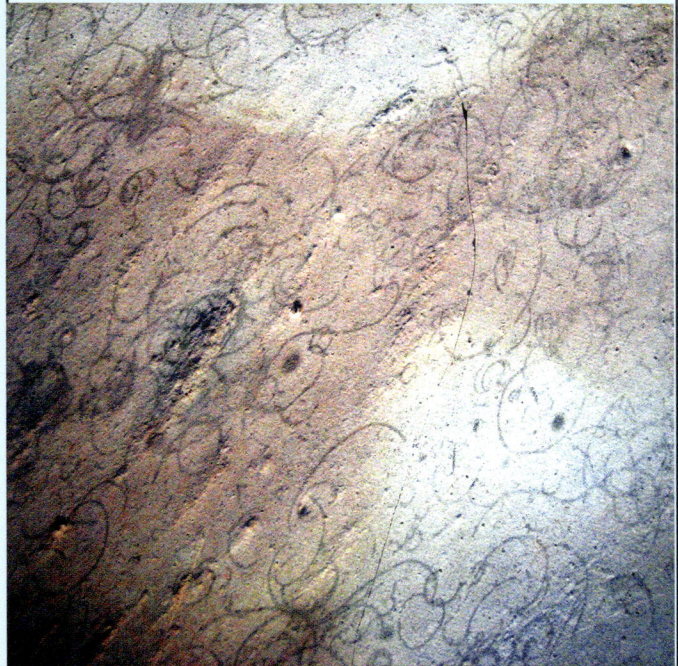

Banded Iron Formation Photographs

Figure 4.13: A BIF specimen with jasper layers that have eroded less than the iron ore layers.

Figure 4.14: A BIF specimen with no plate tectonic folding.

Figure 4.15: A BIF specimen with significant plate tectonic folding.

Figure 4.16: Some BIF specimens contain mustard-yellow limonite, which is a hydrated form of iron.

Figure 4.17: A weathered BIF specimen from the Gitche Gumee Museum's collection.

Figure 4.18: BIF specimen from South Africa.

Figure 4.19: This BIF specimen was found in Ishpeming, Michigan.

Figure 4.20: This 3.0 billion-year-old BIF specimen from Canada contains magnetite.

Figure 4.21: A BIF deposit found in Minnesota.

Figure 4.22: The silica-rich layers in this BIF outcrop are dark in color.

Figure 4.23: This Minnesota BIF cliff, which has been folded by plate tectonics, is 2.69 billion years old.

Breccia

Breccia is one of two types of clastic sedimentary rocks that contain particles larger than .08 inches in diameter (2 mm). The fragments in clastic rocks are easily seen without magnification. When the cemented fragments are angular in shape because they have not been rounded by transport away from their source — they combine to form breccia. When the fragments in a clastic rock are rounded from water transport before being fused, they form conglomerate. Some rocks can be halfway in between. Breccia rocks will be described here; conglomerates will be covered later in the chapter.

The natural cement that cemented rock fragments in breccia specimens may or may not have been the same material as the fragments. The "gluing agent" may have been a chemical precipitate or a sedimentary deposit. The natural cement binding most breccia fragments was most often calcite, silica, or iron oxide. The fragments in breccia can be of any size ranging from huge blocks down to sand and silt. All breccias are grouped into two broad categories, depending on the arrangement of internal components.

Clast Supported: These breccias have angular fragments so numerous they touch each other and dominate the rock. Clast supported breccias have minimal amounts of natural cement filling the small voids between fragments (see Figure 4.24).

Matrix Supported: These breccias have fewer angular fragments more loosely arranged so they do not touch each other. Instead, matrix supported breccias are dominated by the fine-grain matrix or natural cement seams filling the large gaps between and completely surrounding the breccia fragments (see Figure 4.25).

Figure 4.24: This Ontario clast-supported breccia cliff is 2.7 billion years old. It has minimal natural cement between rock fragments.

Figure 4.25: This Oregon matrix-supported breccia has large seams between fragments that filled with natural cement.

Breccia rocks can be made up of a wide array of components. The composition is determined by the source rock from which the breccia formed. The source rock is often used as an adjective when referring to the type of breccia. Sometimes the component fragments are uniform in rock type (monomictic). Examples include sandstone breccia, limestone breccia, granite breccia, chert breccia, and basalt breccia. Other breccias contain fragments from many different types of source rock (polymictic). There are several different geologic environments in which breccia rocks formed. Some are described below.

Sedimentary Breccia: Sedimentary breccias consist of randomly arranged angular to subangular rock fragments that were produced by mechanical weathering and erosion of nearby rocks. This type of breccia can contain a wide array of source rock fragments. In some cases, sedimentary breccias formed when the ceilings of caves collapsed (see Figure 4.26).

Figure 4.26: Sedimentary breccia from a cave collapse in Arkansas.

Tectonic Breccia: This type of breccia formed along geologic faults when continents or blocks of rock slid past each other. The grinding action of the two fault blocks moving against each other broke off pieces of rock to produce the breccia fragments. These fault zones were easily infiltrated by mineral-rich groundwater. Minerals precipitated out of solution and filled the voids between the rock fragments to cement the components together (see Figure 4.27).

Figure 4.27: Tectonic breccia that formed when two large blocks of rocks moved past each other. These breccia fragments are still being fused.

Volcanic Breccia: There are several types of volcanic breccia. The first formed when gases in magma were explosive enough to break igneous rock into fragments that later were re-cemented into breccia. Pyroclastic breccia formed in a matrix of volcanic ash when shattered erupted fragments were cemented together. In the Lake Superior region, the tops of basaltic lava flows also broke up and were later mineralized and cemented together. The natural "gluing" agent in volcanic breccia was either subsequent molten lava that filled in the cracks between fragments, or minerals that precipitated out of circulating fluids (see Figure 4.28).

Impact Breccia: This type of breccia formed from the impact of a large meteorite. The force of the impact shattered rock and deposited fragments throughout the local area. The heat from the impact melted some of the rock which then served to fuse the rock fragments together (see Figure 4.29).

Figure 4.28: Volcanic breccia.

Figure 4.29: This impact breccia formed after a meteorite impact in northern Canada. The specimen is 2.2 inches wide (5.5 cm).

Rock Identification: Breccia

A specimen can be identified as breccia if it has the following characteristics.

Rock Identification Tips	Photographs or other identification information
Breccia rocks have angular rock fragments that were fused together by volcanic material or minerals that precipitated out of solution.	**Figure 4.30:** A cross-polarized microscopic image of Arizona breccia shows fragments cemented together by blue azurite. Field of view is 3.2 inches (8.1 cm). 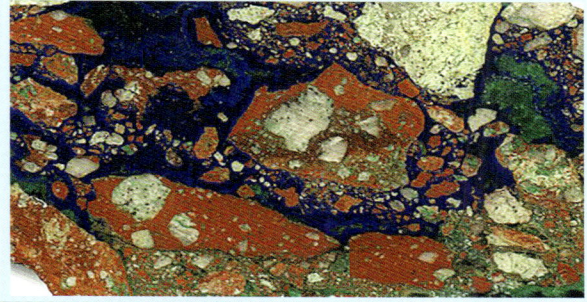
The embedded fragments in breccia specimens must be angular, jagged, and blocky, not rounded as they are in conglomerates.	**Figure 4.31:** A basaltic breccia from a Lake Superior beach. 
The rock fragments in breccia should be easily visible to the naked eye. Otherwise, the properties of this rock type are highly variable. Breccia rocks have a wide array of colors, hardnesses, and other characteristics.	**Figure 4.32:** The irregular shaped fragments in breccia are visible to the naked eye. This jasper breccia is from Romania. 

Rock Identification Tips	Photographs or other identification information
Some breccia specimens are dominated by the natural cement that fused the rock fragments together.	**Figure 4.33:** *A matrix-supported breccia dominated by the fusing cement.* 
Some breccia specimens are dominated by the rock fragments themselves. These breccias have significantly less natural cement.	**Figure 4.34:** *A clast-supported breccia dominated by the rock fragments* 
The hardness of breccia specimens varies significantly, depending on both the embedded rock fragments and the natural cement.	*Because breccia specimens vary in their hardness, it is not useful to use the Mohs Hardness Scale to test these specimens.*
Breccia specimens are opaque.	*Use a bright flashlight to verify the specimen is opaque.*
Breccias can contain fossils, but not often.	*Sometimes fossil fragments are included in breccia rocks.*

Breccia Photographs

Figure 4.35: Sometimes during agate formation, the structure broke apart, fell to the bottom of the vesicle pocket, and became fused as breccia.

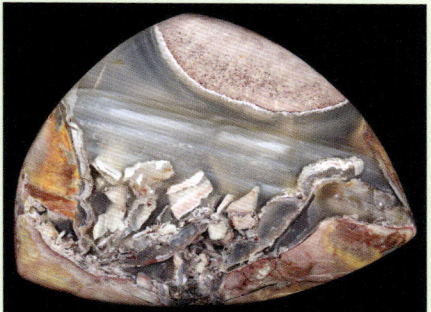

Figure 4.36: This Ohio limestone breccia formed 300 million years ago in a sedimentary environment. Calcite was the natural cement.

Figure 4.37: A clast-supported breccia dominated by the fragments.

Figure 4.38: A tectonic breccia that formed in a geologic fault zone.

Figure 4.39: This basaltic breccia was found on a Lake Superior beach.

Figure 4.40: A matrix-supported breccia dominated by the natural cement.

Figure 4.41: A septarian nodule with brecciated rock fragments.

Figure 4.42: Another basalt breccia.

Figure 4.43: Some breccia specimens can be quite colorful.

Figure 4.44: An igneous breccia that formed at the top of a lava flow in Northern Minnesota.

Figure 4.45: This basaltic breccia was found on a Lake Superior beach.

Figure 4.46: A matrix-supported breccia dominated by the natural cement.

Chert

Chert is a hard, dense sedimentary rock consisting mostly of interlocking microscopic crystals of silica. It is often considered a rock rather than a mineral, even though it is composed almost entirely of microcrystalline silica. It has a hardness of 7.0 on the Mohs Hardness Scale (scratches glass), fractures conchoidally (curved or shell-shaped), and has a smooth surface with a waxy or resinous feel and luster. Chert with hydrated silica and other impurities may be softer with a hardness of 6.5. Unlike quartz, chert is never transparent and unlike translucent chalcedony it is almost always opaque. Chert is sedimentary and therefore often has layers or bands signifying various stages of sedimentary deposition (see Figure 4.47). Banded chert can be distinguished from banded agate by its opaque nature, as compared to the translucency in agate. Chert sometimes contains fossils (see Figure 4.48).

Figure 4.47: Banded chert.

Figure 4.48: Fossiliferous chert.

Chert varies in color but is usually gray, tan, brown, or cream-colored (see Figure 4.49). The peanut butter-colored chert contains limonite, a variety of iron ore. Red varieties are often classified as jasper. Dark varieties that formed in chalk sedimentary deposits are usually called flint. The darker colors resulted from inclusions of mineral matter and organic remains. All three silica varieties — chert, jasper, and flint — consist of microscopic round, granular silica crystals, as compared to the fibrous-shaped microcrystals that make up the varieties of chalcedony. Chert is sometimes considered a lower quality version of flint.

Chert developed in different environments. Like other sedimentary rocks, chert started with the accumulation of particles. In this case, the particles consisted of skeletons from microscopic creatures that spent their lives floating in oceans. These organisms developed their skeletons by secreting either calcium carbonate or silica. When the organisms died, their skeletons sank to the ocean floor, dissolved, and accumulated into a blanket of microscopic sediment called ooze. The microorganisms with silica exoskeletons included sponges, diatoms (single-cell plants, see Figure 4.50), and radiolarians (single-cell animals). When the ooze was mostly calcium carbonate, the sediments hardened into limestone (featured later in this chapter). When the ooze was mostly silica, it crystallized into chert.

Figure 4.49: Common cream-colored chert.

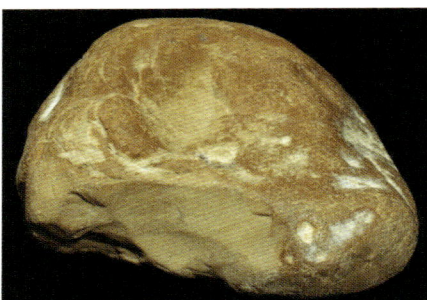

Figure 4.50: When microscopic diatoms with silica exoskeletons die and settle to the seafloor, they chemically change into an ooze, which develops into chert. Most diatoms are between 20 and 200 microns in size (.0008 to .008 inches/.02 to .2 mm).

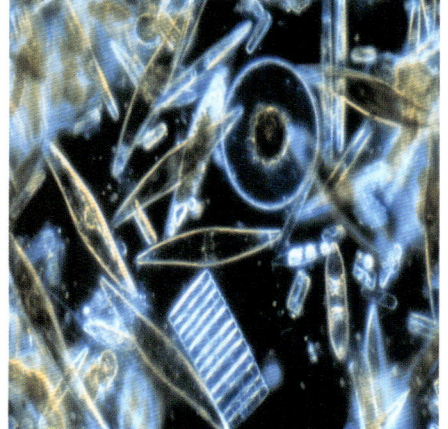

In prehistoric times, chert was used by indigenous people as a raw material for the construction of stone tools. Like obsidian, flint, and similar silica rocks, chert fractures conchoidally allowing it to be knapped into arrowheads, spear points, and other tools (see Figure 4.51). Chert also played another important role during historic times. When a chert stone is struck against an iron-bearing surface, sparks result due to its piezoelectric property. Piezoelectricity is the electric charge that accumulates in certain solid materials in response to an applied mechanical force. This property made both chert and flint excellent tools for starting fires. This characteristic also allowed the use of chert in flintlock firearms, in which a metal plate struck the chert to produce a spark that ignited a small reservoir of black powder, which discharged the firearm. Today, this piezoelectric property is used in quartz watches and in starters on stoves and grills.

Figure 4.51: Chert was used by prehistoric people to make stone tools.

Rock Identification: Chert

A specimen can be identified as chert if it has the following characteristics.

Rock Identification Tips	Photographs or other identification information
Most chert specimens have a waxy luster. Rotate the rocks in either the sun or a bright light to see this luster. **Note:** some weathered chert rocks no longer have a waxy luster.	*Figure 4.52:* Most chert specimens have a waxy luster. 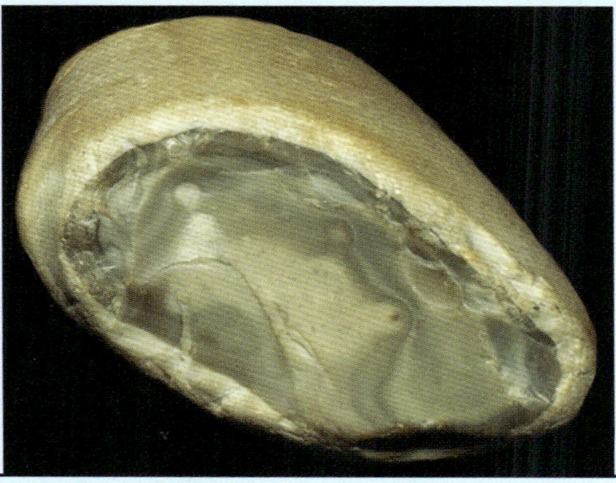
Chert rocks fracture in a conchoidal pattern, shaped somewhat like the curved inside of a seashell.	*Figure 4.53:* Some chert specimens exhibit conchoidal fractures. 
Because of the dense, microcrystalline structure of chert, it usually has a smooth feel and texture.	*Figure 4.54:* Chert specimens have a smooth feel and texture. 

139

Rock Identification Tips	Photographs or other identification information
Some light-colored chert specimens resemble limestone. Pour vinegar on the rock. If it fizzes, it is limestone and not chert. Also, check hardness: limestone will not scratch glass; chert will. Finally, chert has a waxy luster and conchoidal fractures. Limestone does not.	*Figure 4.55: Some chert specimens resemble limestone.* 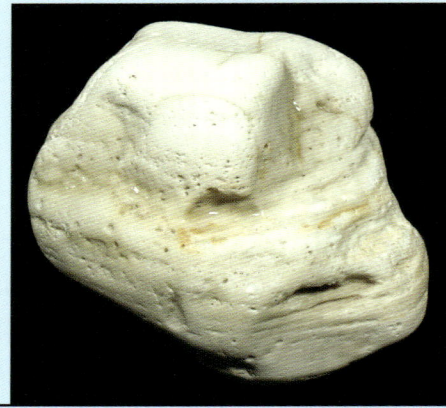
On the Mohs Scale chert has a hardness of 7.0, so it will scratch glass.	*Figure 4.56: Chert will scratch glass.* 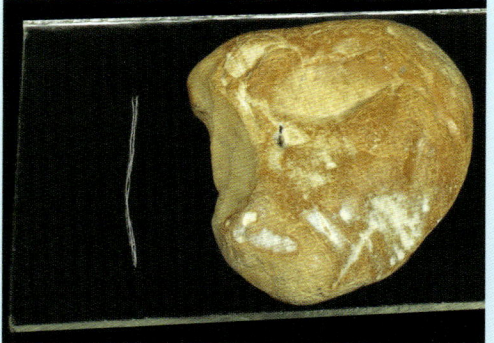
Chert can have banding. Banded chert can be differentiated from agate by checking translucency. Agate is translucent, even if just along the outer edge. Chert is opaque.	*Figure 4.57: Some chert specimens have banding.* 

Rock Identification Tips	Photographs or other identification information
Chert specimens sometimes contain fossils.	**Figure 4.58:** Chert can contain fossils.
Chert specimens are opaque.	**Figure 4.59:** Use a bright flashlight to verify the specimen is opaque.

Chert Photographs

Figure 4.60: This gold-colored chert was found on a Lake Superior beach.

Figure 4.61: Some chert specimens can be two-toned in color.

Figure 4.62: Some chert is not waxy due to impurities or weathering.

Figure 4.63: Layers developed in this chert specimen due to differential erosion.

Figure 4.64: This gray chert rock is from a Lake Superior beach.

Figure 4.65: A multi-colored chert from Indiana.

Figure 4.66: This group of chert stones exhibit waxy luster and conchoidal fractures. Fracturing sometimes creates chert with an irregular shape.

Figure 4.67: A fossiliferous chert wherein silica replaced the organic material, causing a 3D-appearing structure.

Jasper

Jasper is a variety of chert that is often red but can be any number of colors depending on the mineral impurities. It can form in igneous, sedimentary, and metamorphic environments. Jasper is a variety of chert — the microcrystals in both forms of silica are round and granular, as compared to the fibrous shape of chalcedony microcrystals. Like chert, jasper is opaque and does not allow light to pass through.

Jasper can develop in large masses, in amygdaloidal vesicles, as secondary fills in fractures, or as fossil or permineralization replacement, such as in petrified wood (see Figures 4.68 to 4.71).

Figure 4.68: Jasper in a Banded Iron Formation outcrop, located in Ishpeming, Michigan.

Figure 4.69: Jasper in basalt.

Figure 4.70: Jasper secondary fills in basalt.

Figure 4.71: Various types of silica replaced the organic structure in this petrified wood specimen, including jasper.

In the Midcontinent Rift area, most jasper is reddish in color with a waxy feel and appearance (see Figure 4.72). It has a hardness of 7.0 on the Mohs Scale. Jasper pebbles can be found on shorelines as well as in riverbeds, road cuts, gravel pits, and mine dumps. In the Lake Superior region, the predominant red color is derived from iron oxide and distinguishes it from its chert, which is white, tan, or gray. However, chert and jasper can grade into each other. Jasper fractures conchoidally and can be banded. Jasper is opaque, which distinguishes it from red carnelian that is translucent. A jasper specimen streaked with a gray metallic mineral is classified as jaspilite, which is a rock called Banded Iron Formation (see previous section in this chapter).

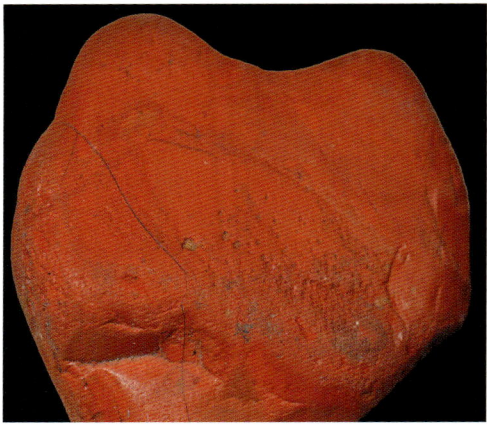

Figure 4.72: Jasper specimen found on a Lake Superior beach.

Rock Identification: Jasper

Worldwide, there are thousands of varieties of jasper that have a wide array of colors. For many jasper varieties, the characteristics used to identify chert (see above section) also apply to jasper. Most jasper specimens do not contain fossils.

Jasper Photographs

Choosing which jasper photographs to include is difficult. There are thousands of jasper varieties in the world, plus other mineral specimens are incorrectly labeled as jasper — they are other types of rock or other types of silica. The photographs included here feature jasper found in the Lake Superior region.

Figure 4.73: *Red jasper on a Lake Superior beach.*

Figure 4.74: *Jasper with quartz-filled secondary seams.*

Figure 4.75: *A group of red jasper specimens.*

Figure 4.76: *This jasper rock has other minerals including gold limonite and white quartz.*

Figure 4.77: *A jasper specimen found on a Lake Superior beach.*

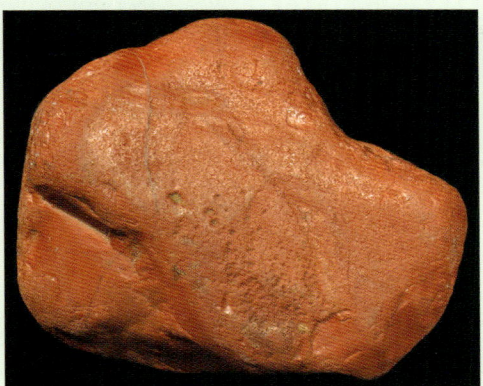

Figure 4.78: *A mixed-silica specimen with mostly jasper.*

Figure 4.79: *A banded jasper.*

Figure 4.80: *Another banded jasper.*

Figure 4.81: This banded jasper has mustard-yellow limonite layers.

Figure 4.82: This jasper specimen exhibits a waxy luster.

Figure 4.83: Banded jasper.

Figure 4.84: Banded jasper containing other silica varieties.

Figure 4.85: Another mixed-silica specimen that includes jasper.

Figure 4.86: This Arizona jasper is similar to rough Lake Superior jasper specimens found inland (not on a beach).

Conglomerate

Conglomerates are sedimentary rocks made up of weathered and rounded to sub-angular fragments larger than .08 inches (2 mm) set in a fine-grained matrix of sand or silt cemented together by calcium carbonate, iron oxide, silica, or clay (see Figure 4.87). To be conglomerate, the fragment components must make up at least 30 percent of the total specimen. Conglomerates formed by the consolidation, compaction, and cementation of gravel. They can be found in sedimentary rock sequences of all ages, but probably make up less than one percent by weight of all sedimentary rocks at the Earth's surface. Conglomerate rocks developed in many different sedimentary environments including river bottoms, alluvial fans, marine settings, and glacial deposition zones. All conglomerates are similar in appearance, although the specific rock pebbles and the natural cement varies widely. At the end of this featured rock section, there is identification information for "generic" conglomerates, as well as some photographs of conglomerates from geographic locations other than the Lake Superior region. Two types of conglomerate found in the Upper Midwest are included as examples and described below.

Figure 4.87: A weathered conglomerate rock.

Copper Harbor Conglomerate

The Copper Harbor Conglomerate is a massive sedimentary formation found on the southwestern part of Isle Royale, along the north shore of the Keweenaw Peninsula, and down the peninsula and into the Porcupine Mountains (see Figure 4.88). It is made up of sediments from volcanic rock weathered and eroded from the Midcontinent Rift as well as from highlands that existed at the time (Huron Mountains and the Mesabi Range). These sediment pebbles, gravel, sand, and fine muds formed in alluvial fan environments at the base of the highlands and along the rift valley (see Figure 4.89). Alluvial fans are fan-shaped deposits of sediments that built up when fast-flowing streams carried sediments down the highland canyons onto flatter, open basins. Over time, these alluvial fans deposits were covered by other layers of sediment. The upper layers compressed the lower layers, compacting the lower sediments closer together. Groundwater circulating through the sediment layers precipitated minerals that cemented the sediments together into Copper Harbor Conglomerate (see Figure 4.90). The Copper Harbor Conglomerate formation ranges between 328 and almost 6,000 feet (100 to 1,800 meters) thick!

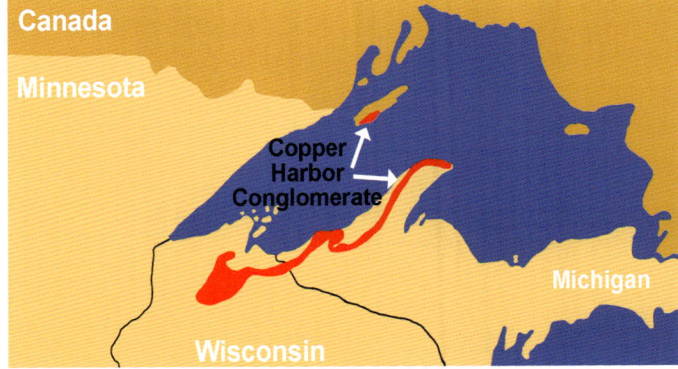

Figure 4.88: Location of Copper Harbor Conglomerate deposit.

Figure 4.89: An arial view of an alluvial fan drainage basin.

Figure 4.90: Copper Harbor Conglomerate rock outcrop.

Copper Harbor Conglomerate Photographs

Figure 4.91: A Copper Harbor Conglomerate boulder.

Figure 4.92: An eroded Copper Harbor Conglomerate outcrop.

Figure 4.93: When Copper Harbor Conglomerate cliffs are exposed, rocks weather out of the outcrop.

Figure 4.94: This Copper Harabor Conglomerate boulder broke off a cliff in the Porcupine Mountain State Park (Michigan).

Figure 4.95: Copper Harbor Conglomerate is exposed at Great Conglomerate Falls in the Black River Pathway Park, located north of Bessimer, Michigan.

Figure 4.96: Copper Harbor Conglomerate deposits are a billion years old!

Puddingstone

A puddingstone is a conglomerate rock containing distinctly rounded pebbles with colors that contrast sharply with the color of the finer-grained matrix. The appearance of this type of conglomerate is thought to resemble Christmas pudding. Puddingstones are found throughout the world. One locality is in the vicinity of the Saint Mary's River, northern Lake Huron, and Ontario, Canada. This conglomerate has several names including Michigan puddingstone, Saint Joseph Island puddingstone, jasper conglomerate, and Drummond Island puddingstone. In these locations, thick beds of conglomerate accumulated between 1.6 to 2.5 billion years ago. Given the age of this puddingstone, some of the deposit has metamorphosed (changed by heat and pressure) to a significant degree. Thus, some of the puddingstone specimens are technically metamorphic rocks, while others did not "cross over the line" and are still considered sedimentary. Given that all the puddingstones found in this area are or were at one time sedimentary, puddingstone is being featured in this chapter.

This local puddingstone variety includes pebbles of red jasper, black chert, hematite, and semi-transparent quartz in a matrix of coarsely grained, light-colored quartzite. For a specimen to be classified as a puddingstone, pebbles visible to the unaided eye must make up between 30 and 90 percent of the components. Scientists conclude this jasper conglomerate formed in alluvial fan environments, from river deposits, or both. These puddingstones can be found in Canada east of Sault Saint Marie, on islands in the Saint Mary's River and northern Lake Huron, and on the beaches and other areas where loose glacial drift gravel can be found. The glaciers spread puddingstone specimens throughout the Lower Peninsula of Michigan and as far south as Ohio and Kentucky.

Puddingstone Photographs

Figure 4.97: A jasper conglomerate puddingstone.

Figure 4.98: Puddingstone slab.

Figure 4.99: This puddingstone was found on a Lake Huron beach.

Figure 4.100: Another jasper conglomerate puddingstone.

Rock Identification: Conglomerate

A specimen can be identified as a conglomerate if it has the following characteristics.

Rock Identification Tips	Photographs or other identification information
Conglomerates are a mixture of round pebbles combined with sand or silt, and cemented together by calcium carbonate, iron oxide, silica, or clay.	*Figure 4.101: This conglomerate slab shows round embedded pebbles fused together into a solid matrix.*
Sometimes the embedded pebbles include both rounded and angular fragments. If most are round, the rock is conglomerate. If most are angular, the rock is breccia.	*Figure 4.102: This puddingstone shows both rounded and angular embedded pebbles.*

Rock Identification Tips	Photographs or other identification information
Conglomerates can be almost any color, or mix of colors, depending on the components.	*Figure 4.103: This colorful conglomerate is from Romania.* 
The grains in conglomerates can vary in size from silt to boulders. The components are not all the same size.	*Figure 4.104: This conglomerate has embedded pebbles of different sizes.* 
Sometimes conglomerates contain fossils or fossiliferous pebbles.	*Figure 4.105: This conglomerate is from a fossil-rich Ohio deposit.* (**Note**: *fossils are not visible.*) 
The hardness of conglomerate rocks varies considerably, depending on the type of embedded pebbles and the hardness of the fusing cement.	*Because the hardness of conglomerate specimens varies, it is not useful to determine the hardness when trying to identify whether a specimen is a conglomerate.*

Rock Identification Tips	Photographs or other identification information
While entire conglomerate specimens are not translucent, some of the embedded pebbles may be translucent.	*Use a bright flashlight to verify that the entire specimen is not translucent.*

Conglomerate Photographs

Figure 4.106: *A conglomerate from northern Minnesota.*

Figure 4.107: *A slab of colorful conglomerate.*

Figure 4.108: *This Ohio conglomerate has pebbles of different sizes.*

Figure 4.109: *A conglomerate from the Gitche Gumee's collection.*

Figure 4.110: *A Brazilian conglomerate.*

Figure 4.111: *A weathered conglomerate from Colorado.*

Sandstone

Sandstone is a clastic rock that makes up around 25 percent of the sedimentary rock on Earth. This rock is composed of small grains between .0024 and .08 inches in size (1/16 and 2 mm) (see Figure 4.112). The vast majority of sandstone rock is comprised of quartz grains, but there are sandstones dominated by feldspar, calcite, or a mixture of different types of rock grains. Like uncemented sands found throughout the world, sandstone can be any color due to the types of grains, but the most common colors are tan, brown, yellow, red, gray, pink, white, and black. The larger coarse-grained sandstones feel the roughest when rubbed, resembling sandpaper. Fine-grained sandstones feel less gritty.

Most pebbles composed of sandstone are hard, dense, and compact due to their erosion-resistant grains as well as the mineral cement (usually quartz) that filled the spaces between grains. Many sandstone deposits have residual porosity between sand grains (see Figure 4.113). Because of this natural porosity, many groundwater aquifers, petroleum fields, and natural gas deposits occur in sandstone deposits. Well-cemented sandstones with less porosity are often mined and used as building material.

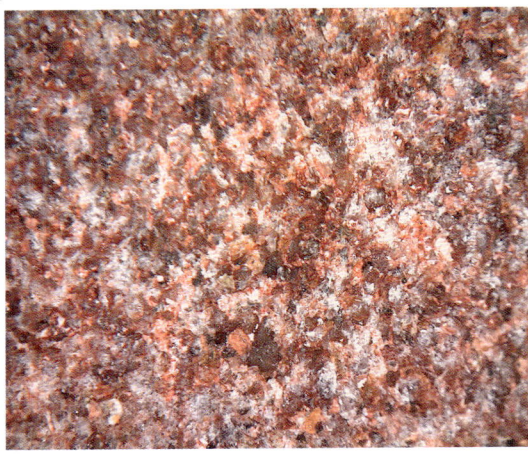

Figure 4.112: Microscopic close-up of sandstone showing the individual mineral grains fused with natural cement.

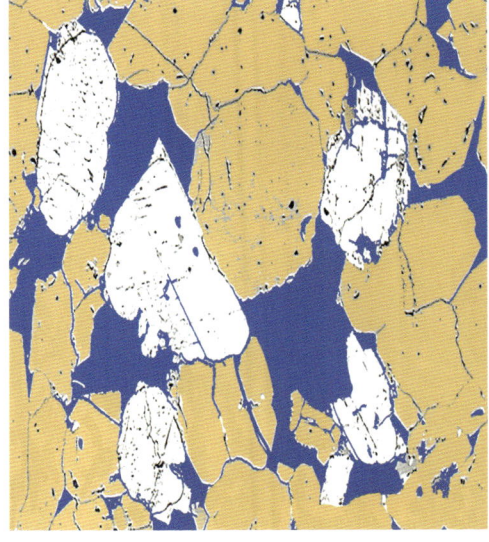

Figure 4.113: This diagram shows how natural porosity between sand grains in sandstone can fill with water (blue areas in diagram).

The compact nature and durability of sandstone, and the vast size of sandstone deposits throughout the world, have created picturesque cliffs and landscapes. For example, much of the rock in the Grand Canyon is sandstone (see Figure 4.114). Sandstone's origin began when bedrock cliffs uplifted by plate tectonic forces were subjected to chemical and physical weathering. The rock broke down into smaller and smaller pieces. Some of the weathered components dissolved into soluble minerals, but the harder grains (e.g., quartz), survived. The eroded and more resistant sand grains were transported from the source area by water, wind, or ice to an area of deposition. Deposition zones included river bottoms, lakes, and oceans. About 80 to 90 percent of the Earth's surface (including ocean bottoms) consist of sedimentary deposition zones. Over time, as the thickness of the deposited sand layers increased, the weight of the top layers compacted the lower layers, fluids were squeezed out from between the grains, and minerals in the fluids precipitated out of solution to cement the grains together. The natural cement that fused the sand grains was usually silica or calcite but may have also been hematite, limonite, feldspar, gypsum, barite, clay minerals, or zeolite minerals.

Figure 4.114: Sandstone cliffs at Supai (Grand Canyon, Arizona).

Worldwide, there are hundreds if not thousands of named sandstones. This section will feature three types of sandstone found in the Lake Superior region. Even within this geographic area, there are other types of sandstone. These three are included to serve as examples.

Graywacke

Graywacke sandstone is a sedimentary rock made up mostly of sand-size grains that were rapidly deposited very near the source rock from which they were weathered. Graywacke formed in deep ocean water near volcanic mountain ranges where underwater landslides and currents transported sediments short distances into a subduction zone or ocean trench. This type of sandstone contains fewer grains made of quartz and more made of feldspar, volcanic rock fragments, silt, and clay. Since there is less quartz and because graywacke rocks are often dark in color, it is sometimes referred to as "dirty sandstone."

The sediment grains in graywacke sandstone deposits include particles of different sizes. The sandstone beds can be from inches to many feet thick and are often separated by thin, dark shale beds. Each graywacke layer was formed during a single submarine landslide event and was deposited over a short period of time from hours to days. The shale beds formed between graywacke deposition events when mud particles slowly settled to the sea floor. The shale layers took hundreds if not thousands of years to form. Within the graywacke layers, the larger and heavier grains settled out first. As the energy in the landslide event decreased, finer and finer particles settled out to the sea floor on top of the larger grains. The larger grains usually make up at least 15 percent of the rock mass by volume. Thus, graywacke formed in the same way as other sedimentary rocks with deposition, burial, compaction, dewatering, and cementation.

Graywacke rocks are mostly gray but can also be brown, yellow, or black. The rock is usually dull-colored and consists of a variety of minerals including quartz, feldspar, calcite, mica, epidote, chlorite, and iron oxides. Sometimes the mineral grains can include chert, slate, gneiss, schist, quartzite, and basalt. Because the mineral grains were not transported far from the source, they are usually angular in shape (see Figure 4.115). The cementing material fusing the mineral grains can be silica, calcite, or clay. Normally graywacke rocks do not contain fossils but may include organic material.

Figure 4.115: *Graywacke rocks have angular mineral grains of different sizes as shown in this microscopic image. Field of view is .2 inches (5 mm).*

Rock Identification: Graywacke

Graywacke is difficult to accurately identify because it can contain so many different types of minerals and rock particle grains. A specimen can be identified as graywacke if it has the following characteristics.

Rock Identification Tips	Photographs or other identification information
Graywacke, as its name implies, is most often gray in color. It can be differentiated from basalt because in most cases graywacke has visible, angular mineral grains, whereas basalt has mineral grains smaller than the human eye can see.	*Figure 4.116:* A collection of graywacke specimens from a beach in England. 
Graywacke can also be green in color if it contains a significant amount of epidote.	*Figure 4.117:* A graywacke rock with green epidote. 
Many graywacke specimens have white quartz veining.	*Figure 4.118:* White quartz veining shows in this graywacke specimen. 
At least 15 percent of the mineral grains in graywacke are visible.	*Figure 4.119:* Graywacke has some visible mineral grains. 

Rock Identification Tips	Photographs or other identification information
Graywacke has a hardness of 6.0 to 7.0 on the Mohs Hardness Scale.	*Using a spare piece of glass, firmly grasp a graywacke specimen and rub it in a straight line on the glass. Wipe any residue off and examine the glass to verify there is a scratch.*
Graywacke does not contain fossils.	*Examine the specimen to verify it does not contain fossils.*
Graywacke is opaque.	*Use a bright flashlight to verify the specimen is opaque.*

Graywacke Photographs

Figure 4.120: *Graywacke formed in horizontal sedimentary beds. This deposit in Minnesota is interbedded with shale and siltstone.*

Figure 4.121: *This graywacke cliff was titled by plate tectonic forces and then scraped by glaciers.*

Figure 4.122: *This close-up image shows angular grains in graywacke.*

Omarolluk (Omar)

Omarolluk, sometimes shortened to omar, are a distinctive type of sandstone consisting of dark sedimentary graywacke stone that have prominent rounded, often deep, hemispherical depressions or holes. Graywacke is a type of sandstone that is dark in color with poorly sorted angular grains of quartz, feldspar, and other rock fragments bound together in a clay matrix (see graywacke information above). These unusual sedimentary rocks are named after the Omarolluk Formation in the Belcher Islands in the southeast corner of Canada's Hudson Bay (see Figure 4.123). The glaciers eroded omars from the Belcher Islands and distributed them in glacial deposits throughout central Canada and the upper Midwest of the U.S. Because scientists know precisely where omars came from, the geography of their distribution indicates the exact movements of the Pleistocene glaciers. Because all omarolluk rocks were glacially transported from the Belcher Islands many thousands of years ago, once these specimens are all taken from the Great Lake beaches and other locales, there will not be a way to replenish the supply until there is another glacier. So, please, either leave them on the beach or only take a couple so some are left for others to enjoy.

Figure 4.123: All omars come from the Belcher Islands, located in Canada's Hudson Bay. They were dragged by glaciers to the southwest into the Great Lakes region.

Omars have a color like igneous basalt rock, but they have at least some mineral grains that are visible. What makes them distinguishable is the hemispherical depressions that are usually between a half inch and three inches in diameter (1.3 to 7.6 cm). The holes are left over pockets formed by concretions that grew in the sediment, became incorporated into the graywacke rock, and then eroded away (see Figure 4.124 to 4.125). A concretion is a compact mass of matter formed by the precipitation of mineral cement within the spaces between sedimentary particles. They are usually ovoid or spherical in shape, although some have more irregular forms. Concretions formed within sedimentary layers that were already deposited early in the burial history of the sediment before the sediments hardened into rock. Over time, the concretions eroded out of the graywacke rock leaving behind the hollow pockets.

Figure 4.124: A concretion nodule is still embedded in this Colorado sandstone.

Figure 4.125: This concretion nodule has eroded out from its sedimentary host rock.

Omarolluk Photographs

Figure 4.126: An omar with two hollow depressions.

Figure 4.127: This omar has secondary filled fracture seams and a hole all the way through the specimen.

Figure 4.128: An omar found on a Lake Superior beach.

Figure 4.129: Another omarolluk found on a Lake Superior beach.

Figure 4.130: This omar is smaller than most at 2.0 inches (5.1 cm).

Figure 4.131: A group of omars.

Jacobsville Sandstone

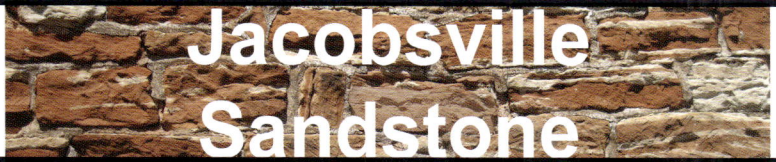

At the end of the Midcontinent rifting period, the weight of the igneous rock filling the rift valley was so extreme that it sank and depressed the Earth's crust. This deepened the rift valley, allowing significant amounts of sand sediments to be deposited by rivers and streams into the depression. The source of these sediments was the highlands that existed south of the rift valley depression (see Figure 4.132). Over time, this mineral-rich sand piled higher and higher, where it eventually became buried by other sediments. Although it did not come close to filling the depression, it did pile sediments thousands of feet thick. Later, these sediments hardened into the beautiful, multi-colored Jacobsville Sandstone.

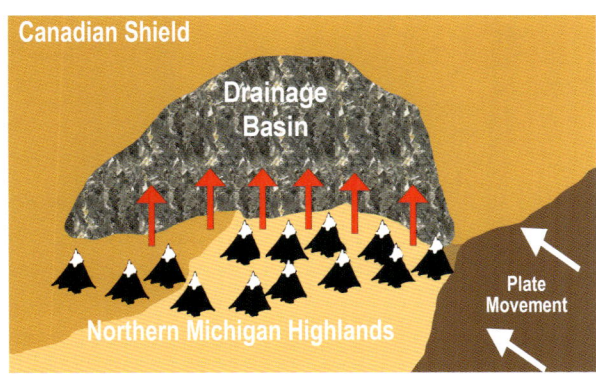

Figure 4.132: The source of sediments for Jacobsville Sandstone was the Northern Michigan Highlands, which have since eroded away. The sediments were deposited into the depression left over from the Midcontinent Rift valley.

Jacobsville Sandstone is an incredibly old sedimentary rock formation that is multi-colored, but is mostly dominated by red. This sandstone is known to be marked with light-colored streaks and spots (see Figure 4.133). The Jacobsville deposit is primarily found along the south shore of Lake Superior but may lie under the entire west end of Lake Superior. The verified deposits in what is now the Upper Peninsula of Michigan are shown in Figure 4.134. This map shows where Jacobsville Sandstone deposits are either exposed, or where they have been verified by well drilling records. The depth of the Jacobsville Sandstone layers varies considerably across its range from 15 feet to over 1,800 feet deep (5 to 550 meters). Due to its color and durability, this sandstone was used as an architectural building stone locally and was also shipped worldwide. There were over thirty Upper Peninsula quarries mining this stone between 1870 and 1915. Buildings throughout the Upper Peninsula, such as at the Quincy Mine in Hancock, Michigan, are made from Jacobsville Sandstone (see Figure 4.135).

Figure 4.133: Jacobsville Sandstone often has multi-colored streaks or spots.

Figure 4.134: The known Jacobsville Sandstone deposit in Michigan's Upper Peninsula is shown in red.

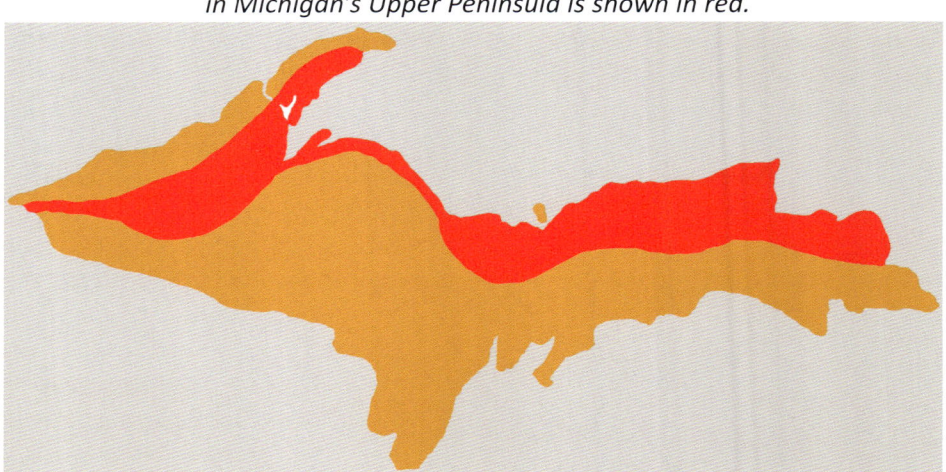

Sediments in Jacobsville Sandstone originated from older rocks (granite and gneiss), which were weathered into sand-sized sediments. Quartz sand grains make up most of this sandstone (up to 75 percent), with feldspar being the next most common mineral (averages 15 percent). The sand was transported and deposited by rivers into the depression that evolved from the Midcontinent Rift

Figure 4.135: Jacobsville Sandstone was quarried and used to make building materials.

Figure 4.136: Many Jacobsville Sandstone specimens have light-colored spots.

valley. As the layers of sediment piled up, the weight of the top layers squeezed and compressed the bottom sediments. As the bottom sediments compacted, they were cemented together by mostly clay and iron oxide, but minerals left behind by water also served as cementing agents (e.g., calcite).

Jacobsville sandstone is most often red but is also mottled with various pinks, whites, and browns. The red color is from iron oxides present in the sediments. After the sands were deposited, ground water moved through pore spaces between the grains of sand. The water contained iron oxide, which circulated through the sedimentary layers and stained the sand grains a rusty red color. There are often light-colored streaks, blotches, and spherical spots in this sandstone that scientists think may have developed in two ways. When acidic fluids circulated across bedding planes or along fractures in the rock, the acid sometimes dissolved and removed iron oxide making these spots and streaks in the sandstone. The second hypothesis proposes that the white spots mark areas where biologic remains were caught in the sediment. This biologic material formed a shield and prevented oxidation and mineral deposition in these spots, leaving light-colored markings (see Figure 4.136).

There is disagreement among scientists about the age of this sandstone, but most believe it formed around a billion years ago. The formation is of terrestrial origin (not marine) and is entirely devoid of fossils. Most of the sand grains in Jacobsville Sandstone range in size from 0.0098 to 0.0197 inches in size (.25 to .50 mm, see Figure 4.137). The sub-rounded shape of the sand grains indicates the transport distance between source and deposition zone was no more than tens of miles.

Figure 4.137: Magnified image of the sand grains in Jacobsville Sandstone.

Rock Identification: Jacobsville Sandstone

A specimen can be identified as Jacobsville sandstone if it has the following characteristics.

Rock Identification Tips	Photographs or other identification information
This sandstone is composed of semi-rounded grains of sand that are all the same size. When you rub a specimen, it feels like sandpaper.	**Figure 4.138:** Jacobsville Sandstone feels gritty like a fine-grit sandpaper. 
Jacobsville Sandstone shows color variations but is mostly red, beige, or brown.	**Figure 4.139:** Jacobsville Sandstone varies in color. This is a large outcrop located along Lake Superior. 
This rock often contains random light-colored streaks and spots.	**Figure 4.140:** A Jacobsville specimen with light-colored streaks. 
Although the sand grains are less than a millimeter in size, it is useful for identification purposes to use a high-powered magnifying glass to verify that the individual sand grains are cemented together with a natural cement.	**Figure 4.141:** This magnified image (x 90) shows how iron-oxide stained some of the sand grains and cemented the grains together. 

Rock Identification Tips	Photographs or other identification information
Some hand-held specimens show sedimentary layers, but layers are more evident in larger Jacobsville deposits.	*Figure 4.142: Jacobsville layers show in this specimen.*  *Figure 4.143: Jacobsville layers are evident in this cliff.* 
Sandstone has a hardness of 7.0 on the Mohs Scale and will scratch glass.	*Using a spare piece of glass, firmly grasp a Jacobsville Sandstone specimen and rub it in a straight line on the glass. Wipe any residue off and examine the glass to verify there is a scratch.*
Although many sandstone rocks do contain fossils, Jacobsville Sandstone does not due to its billion-year-old age.	*Examine the specimen to verify it does not contain fossils.*
Jacobsville Sandstone is opaque.	*Use a bright flashlight to verify the specimen is opaque.*

Jacobsville Sandstone Photographs

Figure 4.144: A pile of Jacobsville Sandstone specimens.

Figure 4.145: A Jacobsville specimen with both light-colored spots and streaks.

Figure 4.146: A Jacobsville deposit on a Lake Superior beach.

Figure 4.147: Another Jacobsville Sandstone beach.

Figure 4.148: Jacobsville rocks dominate this Lake Superior beach.

Figure 4.149: A Jacobsville cliff.

Figure 4.150: Jacobsville specimen. *Figure 4.151:* A Jacobsville cliff along Lake Superior.

Figure 4.152: The exterior of this building is made from Jacobsville.

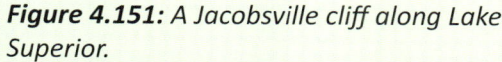

Munising Formation Sandstone

The Munising Formation deposit makes up part of the cliffs in the Pictured Rocks National Lakeshore, located on the south shore of Lake Superior (see Figure 4.153). The cliffs stretch for about 15 miles (24 km) of the 42-mile-long (68 km) park between Munising and Grand Marais, Michigan. The original Munising Formation layers, which were at least 1,700 feet thick (520 m), eroded over time to the white to light gray sandstone layers now rising 50 to 200 feet above Lake Superior. When these sandstone layers were deposited, this area was a complex shoreline/shallow water environment with significant rivers, huge waves, tides, and wind events. Sediments were deposited in the shallow inland sea and nearshore deltas that covered the area. This sandstone can be found in the western part of the Pictured Rocks National Lakeshore along the lake from Munising to Beaver Basin. For a short distance inland from the escarpment, some sandstone cliffs are exposed, especially in the Beaver Basin area.

Figure 4.153: Location of Pictured Rocks National Lakeshore in Michigan.

The Munising Formation layers, which developed around 500 million years ago, are divided into three sections: a basal conglomerate layer, the hard Chapel Rock Sandstone layer, and the crumbly Miners Castle Sandstone layer. These layers lie above the billion-year-old Jacobsville Sandstone deposit. About 500 million years of rock layers between the Jacobsville and Munising Formation deposits no longer exist since they were uplifted and eroded away. Above the brittle Miners Castle Sandstone is the more resistant Au Train Formation, which consists of light brown to white dolomitic sandstone. This harder sandstone forms a caprock that protects the cliffs from erosional forces and is responsible for the existence of several waterfalls in the area. The upper layers of the Pictured Rocks cliffs — and the sedimentary layers deposited above them 0151 contain a significant number of fossils.

The Munising Formation Sandstone specimens found on Lake Superior shorelines tend to be monotone in color (see Figure 4.154), usually tan or beige, as compared to the multi-colored (mostly red) Jacobsville Sandstone. To the touch, Munising Formation Sandstone specimens are grittier and more sandpaper-like than are Jacobsville Sandstone specimens.

Figure 4.154: Most Munising Formation Sandstone specimens are monotone in color.

Rock Identification: Munising Formation Sandstone

A specimen can be identified as Munising Formation Sandstone if it has the following characteristics.

Rock Identification Tips	Photographs or other identification information
This sandstone is composed of rounded grains of sand that are all the same size. When you rub a specimen, it feels like gritty sandpaper.	*Figure 4.155: This magnified image (x 10) of Munising Formation Sandstone shows the texture that makes it feel like gritty sandpaper.*

Rock Identification Tips	Photographs or other identification information
Individual Munising Formation Sandstone specimens tend to be monotone in color. They can be beige, light red, or brown.	***Figure 4.156:*** *Munising Formation Sandstone specimens are monotone in color.* 
Although the sand grains are less than a millimeter in size, it is useful for identification purposes to use a high-powered magnifying glass to verify there are individual sand grains cemented together with a natural cement.	***Figure 4.157:*** *This photo-micrograph image shows how sandstone grains are cemented together with a natural cement.* 
Munising Formation Sandstone deposits show depositional layers. Individual specimens usually do not.	***Figure 4.158:*** *This Munising Formation Sandstone outcrop shows depositional layers.* 
The upper layers of sandstone in the Munising Formation contain some fossils.	*Although most of the Munising Formation sandstone specimens found on the Lake Superior beach do not contain fossils, some do.*
Munising Formation sandstone specimens are opaque.	*Use a bright flashlight to verify the specimen is opaque.*

Munising Formation Sandstone Photographs

Figure 4.159: A beige Munising Formation Sandstone specimen.

Figure 4.160: A brown Munising Formation Sandstone specimen.

Figure 4.161: A light-red Munising Formation Sandstone specimen.

Figure 4.162: A Munising Formation Sandstone cliff in the Pictured Rocks National Lakeshore (Upper Peninsula, Michigan).

Figure 4.163: Munising Formation Sandstone cliff.

Figure 4.164: Another Munising Formation Sandstone cliff.

Figure 4.165: Munising Formation Sandstone cliff.

Figure 4.166: Munising Formation Sandstone depositional layers.

Limestone

Limestone is a sedimentary rock composed of chemical sediments of calcium carbonate derived from the skeletal remains of marine microorganisms (see Figure 4.167). Both plant and animal remains contributed calcium carbonate including corals, clams, mussels, crinoids, protozoa, and others. These marine organisms extracted calcium carbonate from the ocean water to create their shells. Over time, the organisms died, and their biologic remains sank to the ocean bottom. Like other sedimentary rocks, these remains piled up, condensed, and solidified into limestone sedimentary rock. Once the calcium carbonate sediments hardened into limestone, the microscopic grains became too small to be seen with the unaided eye.

Figure 4.167: Limestone formed from the accumulation and compaction of marine organisms with calcium carbonate shells and other types of sediments. Field of view = .02 inches (.5 mm).

Limestone is usually white or gray in color and is usually homogeneous in texture but can have a mixture of whole and fragmented fossils plus mud. If iron oxide was present during formation, specimens may have yellow to brown staining. This sedimentary rock has a hardness of about 3.0 on the Mohs Scale. It can easily be scratched with a knife or a copper penny (minted before 1983). Since limestone specimens consist of more than 50 percent calcium carbonate, it can be identified by carefully dripping vinegar onto the specimen and watching for a chemical reaction. If it is limestone, carbon dioxide bubbles will form.

Limestone formed through a variety of processes. In addition to developing from the shells and other remains of dead sea creatures (bioclastic limestone), limestone also formed from the secretion of marine organisms such as algae and coral (biochemical limestone) or precipitated from the evaporation of mineral-rich water (chemical or inorganic limestones — also known as travertine). Approximately 10 percent of all sedimentary rocks in the Earth's crust is limestone.

Given that the Great Lakes region is in the middle of the North American continent — far from saltwater oceans — one may not expect to find large limestone deposits and marine fossils. At one time, though, this area was a marine environment. During much of the Paleozoic Era (544 to 248 million years ago), much of what is now the Upper Midwest was covered by oceans when this landmass was part of the ancient Laurentian continent (see Figure 4.168). During this geologic period, there was an explosion of life as the number of species in the oceans increased almost exponentially. Nearly all present-day animal groups first appeared during this period. Because the oceans covering this area were shallow, the environment was ideal for life forms to flourish, which resulted in the formation of thick limestone deposits.

Figure 4.168: The Laurentian continent became smaller when a shallow ocean spread over the landmass 500 million years ago. At the time, the continent was in the Southern Hemisphere near the equator, due to plate tectonic movement. Limestone formed in parts of what is now the Upper Midwest of North America where sediments were deposited in these shallow seas.

Limestone is a soft rock, so it is vulnerable to mechanical weathering. Also, because limestone is soluble, it easily weathers and breaks down when it is subjected to slightly acidic rain. Over time, acidic fluids eroded limestone formations and created incredible cave systems. A cave or cavern is a natural hole under the Earth's surface that is large enough for a person to enter. The limestone dissolved by either acid rain or natural acid in groundwater that seeped through bedding planes, faults, or other openings. As acidic waters worked their way through the limestone layers to the ceilings of caves, calcite-rich fluids dripped and formed stalactites and stalagmites (see Figure 4.169).

Figure 4.169: These beautiful formations are in Cave of the Mounds, located west of Madison, Wisconsin

Rock Identification: Limestone

A specimen can be identified as limestone if it has the following characteristics.

Rock Identification Tips	Photographs or other identification information
Limestone is usually white, gray, tan, or yellow, but may contain mineral impurities that stained it red, green, blue, or black.	**Figure 4.170:** A gray limestone rock found on a Lake Superior beach. 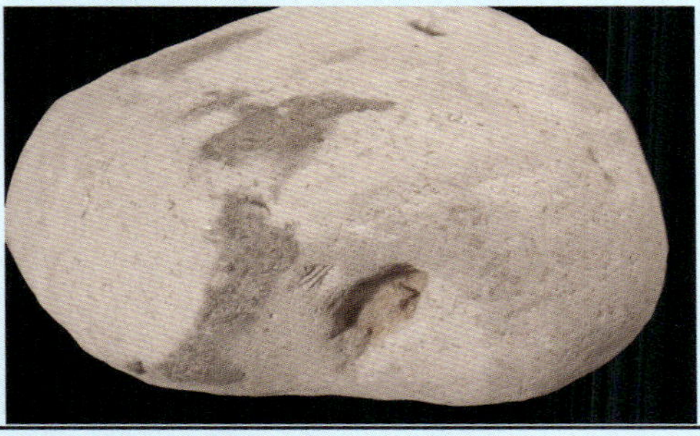
Most limestone specimens have a chalky, smooth texture. Depending on where it formed, the texture can also be sugary, fine grained, or medium grained.	**Figure 4.171:** A chalky, tan limestone rock found on a Lake Superior beach. 
Limestone specimens will fizz when white vinegar or another acid is poured onto the rock's surface. **Note**: working with acids can be dangerous. Follow all safety precautions.	**Figure 4.172:** Check whether the specimen is limestone by carefully dripping an acid (e.g., vinegar) on the rock. If carbon dioxide bubbles form — it is limestone. 

Rock Identification Tips	Photographs or other identification information
Limestone specimens often contain fossils.	*Figure 4.173:* This fossiliferous limestone was found in Michigan's Upper Peninsula. 
Some limestone fossils are fluorescent under UV light.	*Figure 4.174:* Fossils in this limestone found in a river in Michigan's Upper Peninsula fluoresce yellow under 365 nm UV light. 
Limestone is soft with a hardness about 3.0 on the Mohs Hardness Scale.	*If the specimen is limestone, it can be scratched with a steel nail or a piece of copper.*
Limestone is opaque.	*Use a bright flashlight to verify that the limestone specimen is opaque.*

Limestone Photographs

Figure 4.175: A group of limestone pebbles from a Lake Superior beach.

Figure 4.176: This limestone specimen is stained with iron oxide.

Figure 4.177: A typical limestone specimen.

Figure 4.178: The fossil in this limestone is a is a stromatoporaid sponge.

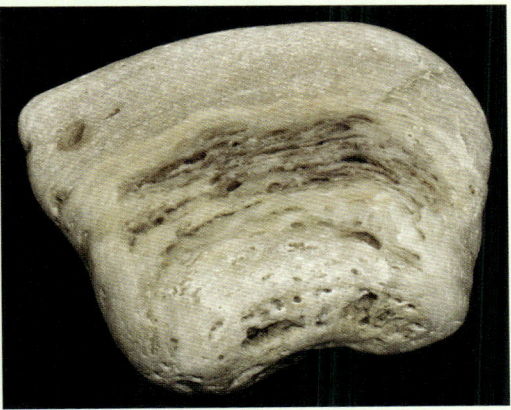

Figure 4.179: This "fossil hash" solidified into limestone.

Figure 4.180: Fossiliferous limestone.

Figure 4.181: A group of rough limestone specimens.

Figure 4.182: This limestone specimen with a trilobite fossil was found in Grand Marais, Michigan.

Figure 4.183: This fossiliferous limestone was found on a riverbank in Grand Marais, Michigan.

Figure 4.184: A close-up of the specimen shown in Figure 4.183.

Figure 4.185: This Indiana oolitic limestone contains rounded calcite grains that formed from being rolled around by waves.

Figure 4.186: Most of the rock on Michigan's Mackinac Island is limestone, including Arch Rock.

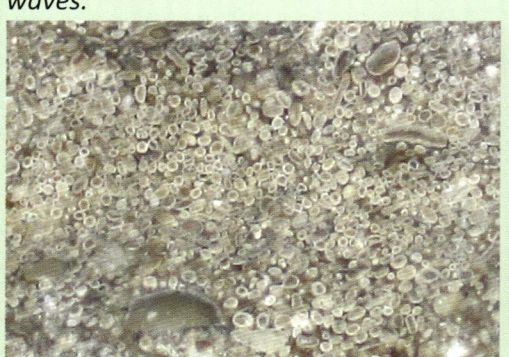

Shale

Shale formed when muddy or clay sediments hardened into rock. Shale is laminated, meaning the rock consists of many thin layers that were bound together (see Figure 4.187). because clay particles are extremely small (< 0.002 mm in diameter), they tended to be transported farthest from their source rock into deep, slow moving, quiet waters. Clay minerals, which are the dominant sediment found on Earth, originate from the weathering of feldspars and other silicate minerals. Much of the seafloor on the bottom of the world's oceans is covered with clay deposits. Shale consists of at least 30 percent clay particles, along with tiny grains of quartz, pyrite, calcite, mica, and organic material. It is classified based on its mineral content. For example, siliceous shale is dominate by silica, calcareous shale by calcite, hematitic shale by iron oxide, etc.

Figure 4.187: Shale formed from sedimentation and compaction of mud and clay, loped into laminations (thin layers).

Shale rocks have a uniform texture and often can be easily split into individual plates along bedding plains (called fissility). Shale has a hardness of between 2.0 and 3.0 on the Mohs Hardness Scale, so it will not scratch glass but can be scratched with a nail, and sometimes with a fingernail. Although it can form thin layers, it is most often found in large, layered masses (see Figure 4.188). Shale is the most common sedimentary rock on our planet making up about 70 percent of the sedimentary rock found in the Earth's crust. Sometimes animal tracks, fossils, or even imprints of raindrops are preserved in shale layers (see Figure 4.189).

Figure 4.188: Shale can form in large masses.

The color of shale depends on its composition. Shale with a higher organic (carbon) content tends to be darker in color – either gray or black. In anoxic environments lacking in oxygen (e.g., buried in thick mud), organic material that did not decay or become fossilized instead formed a dark sediment, rich in organic matter (see Figure 4.190). The presence of iron compounds produced shale that is red, brown, blue, black, green, or purple. Shale containing a lot of calcite tends to be pale gray or yellow.

Figure 4.189: These plant fossils are preserved in Illinois shale deposits.

One common variety of shale found in the western Upper Peninsula of Michigan is Nonesuch Shale. In Figure 4.191, Nonesuch Shale is well exposed in the lower Presque Isle River, located north of Ontonagon, Michigan in the Porcupine Mountains State Park. The rounded pits in the shale are potholes formed by gravel sediments swirling in eddies within the river currents. The gravel-sand mixture deepened pre-existing bedrock depressions.

Figure 4.190: This dark shale was rich in organic matter during formation.

Figure 4.191: Gravel and sand rotated in river eddies and carved out potholes in Nonesuch Shale.

Rock Identification: Shale

A specimen can be identified as shale if it has the following characteristics.

Rock Identification Tips	Photographs or other identification information
Shale is usually black, gray, or brown, but it can also be tinted red, green, or blue depending on its mineral content.	**Figures 4.192 to 4.194:** *Shale colors vary. Here are black, gray, and brownish/red varieties.*   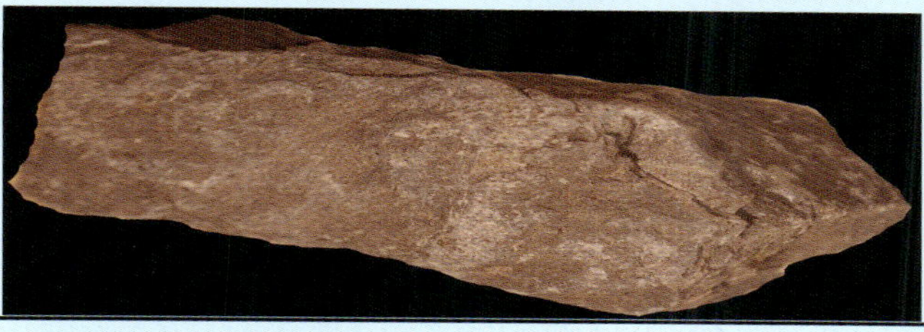
Shale is fine-grained, so the mineral crystal grains cannot be seen without magnification.	**Figure 4.195:** *Shale is extremely fine-grained.* 
Shale is smooth to the touch.	**Figure 4.196:** *Shale slabs have a smooth surface, especially those that have been exposed to the elements.* 

Rock Identification Tips	Photographs or other identification information
Shale can be split into horizontal slabs.	**Figures 4.197:** Because shale can easily be split into horizontal slabs, this rock has been used by many cultures as a construction material. 
If an edge of a shale specimen is dipped in water and dragged along a smooth surface, it will leave a muddy streak. This characteristic helps differentiate shale from slate.	**Figure 4.198:** Wet shale leaves a muddy streak. 
When moistened, shale usually *smells* like wet mud. This characteristic helps to differentiate shale from slate.	**Figure 4.199:** Moisten the specimen and verify it smells like mud. 
Shale has a hardness of between 2.0 and 3.0 on the Mohs Hardness Scale. Shale specimens can be scratched with a nail, and sometimes with a fingernail.	**Figure 4.200:** Use a nail to verify the specimen can be scratched. 

Rock Identification Tips	Photographs or other identification information
Shale often contains fossils. Sometimes to find a fossil, shale specimens must be split into slabs.	**Figures 4.201:** This Ohio shark fossil is from Devonian Period shale (around 400 million years old).
Shale is opaque.	Use a bright flashlight to verify the specimen is opaque.

Shale Photographs

Figure 4.202: Shale layers are shown in this microscopic image.

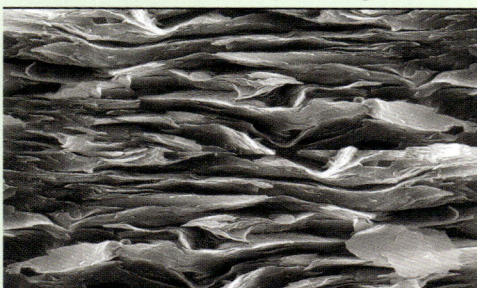

Figure 4.203: Nonesuch Shale layers in the Porcupine Mountain State Park (Michigan).

Figure 4.204: A shale specimen found on a Lake Superior beach.

Figure 4.205: This complex mass of fossil burrows weathered and eroded from Ohio shale.

Figure 4.206: Shale layers.

Figure 4.207: Shale layers showing various colors.

Figure 4.208: An Indiana shale quarry.

Figure 4.209: This waterfall cascades over Nonesuch Shale in Michigan's Porcupine Mountain State Park.

Figure 4.210: Weathered shale layers.

Figure 4.211: This shale specimen has a magnificently-formed ammonite fossil.

Chapter 5: Metamorphic Rocks

Note: For a list of the Featured Rocks included in this chapter, please go to page 189.

Geologic Background

Metamorphic rocks formed when one type of rock was transformed into another type because the source rock was changed by high temperature and pressure. To create metamorphic rock, the source material had to remain solid and not completely melt. If there was too much heat and pressure, the source rock would have melted into magma, which would then have re-crystallized into igneous rock. Different metamorphic minerals formed at different temperatures and pressures. In some cases, completely new rocks developed. In other cases, crystal structure or texture in the source rock was altered. The source rock may have been igneous, sedimentary, or even another metamorphic rock.

There are three main types of metamorphism: burial, regional, and contact. Burial metamorphism occurs when rocks are deeply buried, at depths of below 6,500 feet (2,000 meters). Burial metamorphism commonly occurs in sedimentary basins, where rocks are buried deeply by overlying sediments (see Figure 5.1). Increase of temperature with depth in combination with a rise in confining pressure produces low-grade metamorphic rocks. On average, the temperature below the Earth's surface increases about 1°F per 70 feet of depth (2.5°C per 100 m). Regional metamorphism occurs when changes take place in large

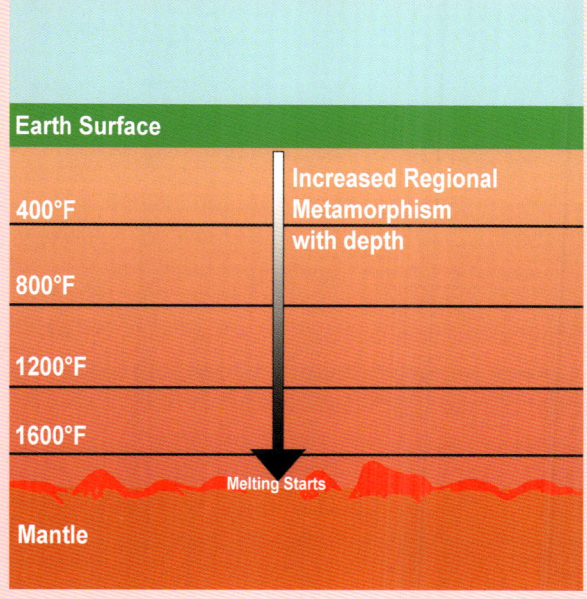

Figure 5.1: Burial metamorphism is caused by the increase in temperature and pressure at depth below the Earth's surface.

masses of rock over a wide geographic area. Regional metamorphism takes place when source rock is subjected to increased temperature and pressure over a large area, and is often located in mountain ranges created by plate tectonic forces (see Figure 5.2). Not only did these plate tectonic forces cause heat and pressure, which metamorphoses rock, but sometimes these massive continental collisions forced layers of source rock to fold. The junction between the two colliding plates is called a plate boundary. On either side of the plate boundary, there are fold-thrust belts. Metamorphic rock forms in the fold-thrust belts on either side of the plate boundary. Contact metamorphism occurs in rock exposed to high temperature and low pressure, such as when hot magma intrudes into or lava flows over pre-existing rock. Oftentimes, this type of metamorphism reorganizes the mineral grains into larger crystals with a courser texture. The amount crystals that changed in the source rock decreases with the distance away from the heat of the magma or lava (see Figure 5.3).

Figure 5.2: Regional metamorphism is caused by the collision of tectonic plates.

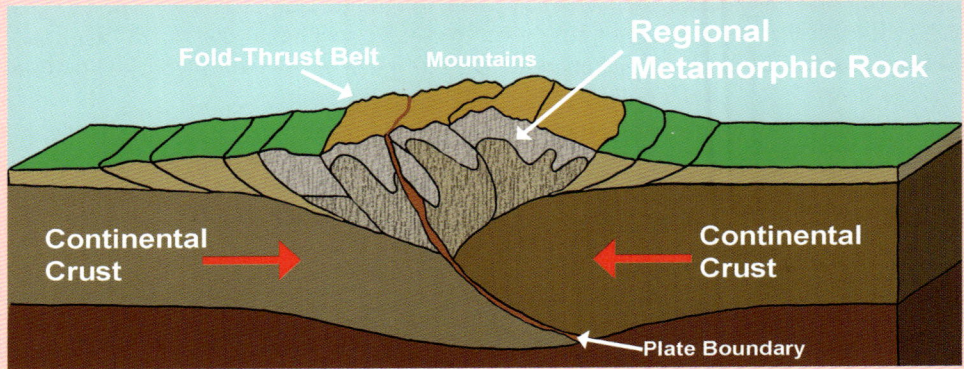

Metamorphic rock makes up around 25 percent of the Earth's crust and covers around 12 percent of the land surface. The study of metamorphic rocks now exposed at the Earth's surface following erosion and uplift provides information about the metamorphic conditions that exist in our planet's interior. Metamorphic rocks contain more types of minerals than the number found in either sedimentary or igneous rocks. Nearly all the minerals found in igneous and sedimentary rocks are present in metamorphic rocks. These minerals formed during the crystallization and solidification of the source rock. Because some of these minerals have high melting points and were stable at the temperatures and pressures that occurred during metamorphism, they remained chemically unchanged when they became incorporated into metamor-

Figure 5.3: Contact metamorphism occurs when molten magma intrudes into or lava flows over pre-existing rock.

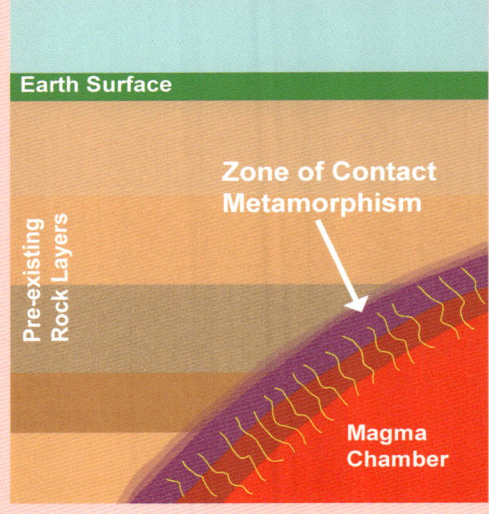

phic rocks. These minerals include olivine, hornblende, mica, feldspar, and quartz. In addition, other minerals uncommon or nonexistent in igneous and sedimentary rocks formed during metamorphism. These are known as index minerals and include those minerals commonly found in metamorphic rocks (e.g. garnet, see Figure 5.4). Other index minerals include chlorite, epidote, staurolite, kyanite, and sillimanite. However, these minerals can also be found in igneous and sedimentary rocks.

Figure 5.4: Garnet schist.

The pressures and temperatures that formed metamorphic rocks created two different types of texture: foliated and non-foliated. Foliated rocks include those deformed and recrystallized along one plane. The mineral grains in foliated rock reorganized into flat striated zones with their longest axes arranged parallel to one another. This realignment of crystal grains allows some metamorphic rock specimens to be easily split in one direction following this long axis of crystal arrangement. One example of a foliated metamorphic rock is slate that metamorphosed most of the time from shale (see Figure 5.5). Gneiss is another foliated rock that metamorphosed from either igneous or sedimentary rock (see Figure 5.6). Non-foliated metamorphic rocks do not have this horizontal arrangement of mineral grains. Marble, for example, metamorphosed from beds of limestone and is a common, crystalline-textured metamorphic rock composed of calcite (see Figure 5.7). Marble varies greatly in color and crystal

Figure 5.5: Slate is foliated with crystal grains arranged parallel along a single directional plane, as shown in this Cornwall, England cliff.

Figure 5.6: Gneiss is also a foliated metamorphic rock.

size, and can be identified by its crystalline texture, by bubbling when subjected to acid, and by not scratching glass (marble has a hardness of 3.0). Marble's uniform texture allows it to be used as a material for sculpture and architecture. Another example of a non-foliated metamorphic rock is quartzite, which developed from sandstones when metamorphism converted sand grains into a mosaic of small, close-fitting crystals of quartz (see Figure 5.8). Both slate and gneiss are featured in this chapter because specimens are readily identifiable. Because marble and quartzite have a wide array of appearances and are harder to identify, they are not featured in this chapter.

Figure 5.7: Marble is a metamorphic rock that does not have horizontal, foliated structure.

Figure 5.8: Quartzite is another non-foliated metamorphic rock.

Metamorphic rocks are identified by their mineral composition and their texture. However, this is challenging because different amounts of heat and pressure applied to the same source rock created metamorphic rocks of the same type that have a varied appearance. Also, a certain type of metamorphic rock may have originated from different source rocks, which varied its appearance. Like other rocks, significant field experience and sophisticated laboratory equipment are needed to definitively identify many metamorphic rocks.

Featured Rocks

In this chapter the metamorphic rocks listed below will be featured because they have characteristics that make them easier to identify. The metamorphic rocks include:

> **Key**
> - Featured rock (includes rock identification information and photographs).

- **Slate** (page 189): A fine-grained foliated metamorphic rock;
- **Schist** (page 195): A foliated metamorphic rock containing more than 50 percent medium to large flat crystal grains;
- **Gneiss** (page 202): A coarse-grained foliated metamorphic with crystal grains loosely arranged in parallel layers or bands;
- **Unakite** (page 208): A multi-colored metamorphic rock that formed when granite was altered by hydrothermal fluids.

Slate is a very fine-grained, low-grade metamorphic rock that usually formed from the alteration of shale. The mineral grains in slate cannot be seen by the unaided eye. Of all metamorphic rocks, slate can be most easily split into layers along the parallel foliated structure. Slate's characteristic to easily split into slabs allows it to be used to make roofing tiles, flagstones, billiard tabletops, flooring, tombstones, commemorative tablets, and blackboards. This rock is especially well-suited for roofing because it does not absorb water and can withstand freezing temperatures (see Figure 5.9).

Slate has a dull luster and is usually gray, green, brown, reddish purple, or blue in color. However, some slate specimens can appear "wet" when viewed in a bright light. The color of slate is determined by the amount and type of iron oxide and organic material present when the rock formed.

Figure 5.9: Slate formed in horizontal foliated layers that allow it to be easily split to make roofing tiles.

One way to identify slate is to strike it with another stone or carefully drop it onto a hard surface: slate makes a ringing "clink" sound.

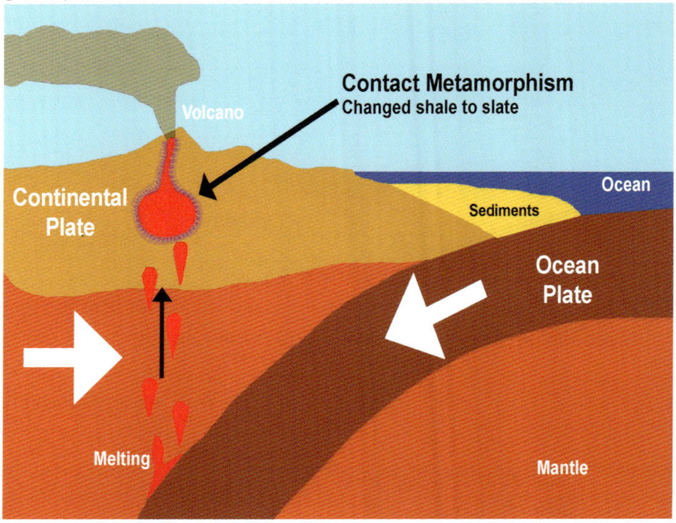

Most slate formed at convergent plate boundaries when ocean crust subducted below continental crust due to plate tectonic forces (see Figure 5.10). In the former sedimentary rock subjected to these forces, heat and chemical activity transformed clay minerals in shale into platy minerals such as muscovite mica, biotite mica, or chlorite. The directed pressure reorganized the clay minerals from their random orientation in shale into a parallel arrangement. The metamorphic forces caused the platy minerals to line up with their long axes parallel to each other, but perpendicular to the direction of the plate tectonic compressive force. This transformation of minerals marks the point in the rock's history when it was no longer sedimentary and instead became the low-grade metamorphic rock, slate. If slate rock was subjected to additional heat and pressure, it would have further metamorphosed into phyllite (see Figure 5.11). Phyllite is a low- to intermediate-grade, foliated metamorphic rock. Phyllite is composed of microcrystalline mica and is usually light gray, dark gray, or black in color. It has a slightly shiny luster under a bright light and often has wrinkled or crinkled foliation planes.

Figure 5.10: Shale metamorphosed into slate at convergent plate boundaries.

Figure 5.11: Phyllite formed from the metamorphism of slate. This specimen is from Wyoming.

The differentiation between slate and phyllite can be difficult, especially because individual specimen grade along the entire continuum between these two rock types. There is no easy way to quantitatively decide when one type ends and the other begins. The individual mineral grains in both slate and phyllite are not visible to the unaided eye, but the platy mineral grains are large enough in phyllite to impart a silky sheen to the surfaces of the rock, which is different than the dull appearance of slate. Because of the difficulty in correctly identifying phyllite, this rock will not be featured.

Problems in correctly identifying slate is increased by the inconsistent historical use of the names "shale" and "slate." In some industries, the terms are used interchangeably. This confusion partially arises from the fact that shale is progressively converted into slate. Since the metamorphic process produced a series of rocks grading from shale to slate (and then beyond to phyllite, schist, and gneiss), a specimen can be identified anywhere along this continuum. There are no convenient definitive lines of delineation.

Rock Identification: Slate

A specimen can be identified as slate if it has the following characteristics.

Rock Identification Tips	Photographs or other identification information
The mineral crystals in fine-grained slate cannot be seen without magnification.	**Figure 5.12:** The mineral grains in slate cannot be seen by the unaided eye. 
Although slate is usually gray, it can occur in a variety of colors depending on its components. The most common colors are green, brown, reddish purple, and blue.	**Figure 5.13:** A green-colored slate specimen. 

Rock Identification Tips	Photographs or other identification information
Slate specimens are smooth to the touch.	*Figure 5.14:* This gray colored slate rock is smooth to the touch. 
Slate can easily be broken into thin sheets because of its foliation.	*Figure 5.15:* Slate can easily be split into thin sheets. 
Slate is a medium-hard rock measuring between 2.5 and 4.0 on the Mohs Hardness Scale. Slate will not scratch glass but can be scratched with a knife.	*Figure 5.16:* Slate can be scratched with a knife. 
Slate can have a wet-like appearance when exposed to either the Sun or a bright light. Otherwise, slate has a dull appearance.	*Figure 5.17:* Slate can have a wet-like appearance under a bright light. 

Rock Identification Tips	Photographs or other identification information
Because slate was formed under low heat and pressure conditions, especially compared to other metamorphic rocks, some fossils can be found in slate, including remains of delicate organisms.	*Figure 5.18:* A brittle star fossil in German slate.
When slate is struck with another rock or dropped onto a hard surface, it makes a distinctive "clink" sound.	Strike the specimen against another rock to determine if it makes a "clink" sound.
Slate is opaque.	*Use a bright flashlight to verify the specimen is opaque.*

Slate Photographs

Figure 5.19: A slate specimen from the western Upper Peninsula (Michigan).

Figure 5.20: Two slate specimens.

Figure 5.21: This slate deposit is stained red with iron oxide.

Figure 5.22: A slate deposit in northern Minnesota.

Figure 5.23: A pile of slate.

Figure 5.24: This pile of slate has been weathered.

Figure 5.25: A slate outcrop that has been tilted up by plate tectonic forces.

Figure 5.26: *Slate surface close-up.*

Figure 5.27: *A blue-gray colored slate.*

Schist

Schist is a metamorphic rock that consists of medium- to coarse-grained mineral crystals visible to the eye or with a hand lens. The mineral grains have a highly preferred orientation. More than 50 percent of the crystal grains in schist are needle-like or platy minerals that lie with their long direction parallel (e.g., hornblende and mica). Often schist rocks can have layers of different minerals: quartz or feldspar layers may lie between hornblende and mica layers. The minerals in schist vary, which considerably changes the appearance of different specimens. The predominate minerals or larger embedded crystals are used to describe the type of schist (e.g., mica-schist, garnet-schist, chlorite-schist).

A rock does not need a specific mineral composition to be called "schist." It only needs to contain enough large platy metamorphic minerals arranged in alignment to exhibit distinct foliation layers. This texture allows the rock to be broken into slabs along the alignment direction of the platy mineral grains (see Figure 5.28). However, schist does not break into slabs as easily as does slate. Another characteristic of schist is that the larger crystal grains reflect light causing schist to have a shiny luster.

Figure 5.28: *A slab of schist.*

Schist is one of the most widespread rock types in the continental crust. Most schist metamorphosed from slate subjected to additional metamorphic forces that caused the mineral grains (especially mica) to grow and elongate (see Figure 5.29). When additional heat, pressure, and chemical activity continued, schist metamorphosed into gneiss (see next featured rock). In rare cases the platy metamorphic minerals in schist did not originate with shale, but instead from basalt or other rock sources.

Figure 5.29: A cross-polarized microscopic image of schist showing the parallel arrangement of large, platy crystal grains. Field of view is .02 inches (.5 mm).

Rock Identification: Schist

A specimen can be identified as schist if it has the following characteristics.

Rock Identification Tips	Photographs or other identification information
Schist and slate are comparable because they are made from similar minerals, but schist was subjected to more severe metamorphic forces. The main way to differentiating these varieties is that in most cases the individual crystal grains can be seen in schist without magnification, whereas they are not visible to the unaided eye in slate.	*Figure 5.30: The mineral grains in schist can be seen by the unaided eye.* 

Rock Identification Tips	Photographs or other identification information
Schist can be split into slabbed layers, but not as easily as slate. The foliated layers of schist can be seen when viewing a specimen from its side.	*Figure 5.31:* The layering of schist is shown in this French outcrop. The schist layers are gray, which are interbedded with quartzite (yellow/orange). This rock outcrop was tilted up by plate tectonic forces. 
Schist specimens often have large flat mica flakes.	*Figure 5.32:* Schist has large mica crystal grains that appear as "flakes." Field of view is around two inches (5.1 cm). 
There may be quartz layers between the mica layers in schist.	*Figure 5.33:* Schist can have quartz layers between mica layers. The field of view is two inches (5.1 cm). 

Rock Identification Tips	Photographs or other identification information
Schist specimens may have large mineral crystals embedded in the layers.	*Figure 5.34:* This schist has blue kyanite crystals. 
The layers in schist may be somewhat wavy, but not as wavy as in gneiss.	*Figure 5.35:* Schist foliated layers can sometimes have a slight "wavy" appearance. 
Large crystal surfaces in schist reflect lots of light, which gives it a shiny appearance.	*Figure 5.36:* Schist often has a shiny appearance, due to the large number of mica crystals. This specimen is from Vermont. 

Rock Identification Tips	Photographs or other identification information
Schist can be differentiated from gneiss because it can be split into layers; gneiss cannot. Both have mineral grains that can be seen with the naked eye, but the grains in gneiss are larger and arranged into obvious dark and light bands. Schist may have layers, but they are not aligned into alternating dark and light bands.	*Figures 5.37 and 5.38: Schist can be differentiated from gneiss by the patterns of layers. Most schist specimens do not have dramatic alternating light and dark layers; gneiss specimens do. Schist is on the left; gneiss is on the right.* 
Most Schist specimens have a hardness of between 4.0 and 5.0 on the Mohs Scale, based on the combined hardness of its mineral components. It will not scratch glass unless it has quartz, but schist can be scratched with a knife blade.	*Figure 5.39: Schist specimens can be scratched with a knife blade.* 
When schist is scratched with a knife, it sometimes releases a bad smell.	*Figure 5.40: When scratched, a smell may be released.* 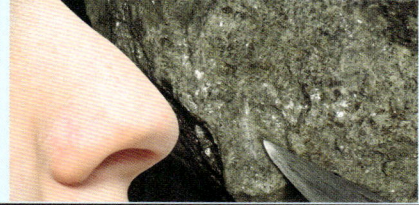
Schist rocks do not contain fossils.	*Examine the specimen to verify it does not contain fossils.*
Schist specimens are opaque.	*Use a bright flashlight to verify the specimen is opaque. Note: embedded crystals may be translucent.*

Schist Photographs

Figure 5.41: This garnet-mica schist cross-polarized microscopic image shows garnet (black), biotite (orange), muscovite (blue), and quartz (gray).

Figure 5.42: This schist specimen was found on a Lake Superior beach.

Figure 5.43: Another schist from a Lake Superior beach.

Figure 5.44: This garnet-chlorite schist is from Tennessee.

Figure 5.45: This 1.7 billion-year-old muscovite-mica schist is from South Dakota.

Figure 5.46: Another schist from a Lake Superior beach.

Figure 5.47: *A chlorite-garnet schist from Alabama.*

Figure 5.48: *This Maryland chlorite schist has black magnetite crystals.*

Figure 5.49: *A group of mica schist with reflective surfaces.*

Figure 5.50: *Foliated layers are seen in this talc schist when viewed from the side.*

Figure 5.51: *This spectacular tourmaline-mica schist is from South Dakota.*

Figure 5.52: *Another mica schist.*

Gneiss

Gneiss (pronounced "nice") is a foliated, high-grade metamorphic rock identified by its horizontal arrangement of dark and light, loosely arranged bands. Some layers in gneiss contain granular crystals with an interlocking structure, while other bands contain platy or elongated minerals with parallel orientation in one direction. It is this loosely banded appearance and texture — rather than mineral composition — that defines gneiss. A significant portion of gneiss rock in the Earth's crust formed during mountain building events where the source rock was subjected to high pressures and temperatures generated by depth of burial, closeness to magma intrusions or plate tectonic forces. Gneiss is so abundant in the Earth's crust that if you drill anywhere on the surface, you will eventually strike gneiss.

Gneiss rocks consist of mostly feldspar, quartz, and mica, but they can also have smaller amounts of hornblende, pyroxene, and biotite mica. Although it may appear that the mineral grains in gneiss are arranged into bands, when examined closely it is clear these mineral streaks are not well-defined and show gaps between the crystals (see Figures 5.53 and 5.54). Gneiss often metamorphosed from granite, but also may have formed from other igneous, sedimentary, or metamorphic source rocks. The alternating dark and light layers, called compositional banding, consist of different minerals. The darker bands contain minerals such as hornblende and biotite mica. The lighter bands contain lighter minerals such as quartz and feldspar.

Gneiss developed in source rock between six and twelve miles (10 to 20 kilometers) below the Earth's surface. The pressures at this depth exceeded 14,500 pounds per square inch (10 kilobars), with temperatures greater

Figure 5.53: The mineral grains in gneiss are horizontally arranged into loose, poorly organized striations.

Figure 5.54: This gneiss specimen shows loosely arranged, striated mineral grains.

than 600°F (320°C). Gneiss can look similar to granite, but in granite the crystal grains are randomly arranged, whereas the mineral components in gneiss have an obvious parallel arrangement. Gneiss can be distinguished from schist by its larger mineral crystals, more obvious parallel crystal arrangement, and because it has poor cleavage and does not break along planes of foliation. The lack of cleavage in gneiss results because the layers are not well developed: less than 50 percent of its crystal grains are aligned in well-defined layers. Because of the coarseness of the structure, the layers are sub-parallel, often discontinuous, and do not have a constant thickness. This unique structure in gneiss rocks is called gneissic banding, which consist of layers that are thicker than those displayed in other foliated metamorphic rocks. When hammered, gneiss breaks in a random, blocky fashion like granite and gabbro, and not at all like the more closely related metamorphic rocks like schist that breaks along horizontal planes (see Figure 5.61 in the **Rock Identification** section below).

In some cases, gneiss metamorphosed from schist, which itself metamorphosed from slate which metamorphosed from sedimentary shale. To form gneiss, the original rock had to be subjected to high pressure and have sufficient time to allow large crystal grains to slowly develop. During this transformation, clay particles in shale transformed into micas and increased in size. Finally, the platy micas recrystallized into granular minerals. The appearance of granular minerals is what marks the transition into gneiss.

Some gneiss deposits are among the oldest rocks on Earth. Few rocks on our planet date beyond 3.7 billion years. In 1999 geologists found what they thought was the oldest rock outcropping on Earth. In Canada's Northwest Territories, Acasta gneiss was dated at 4.031 billion years old (see Figure 5.55). Then, in 2008, they discovered even older rock in the Nuvvuagittuq Greenstone Belt on the east coast of Hudson Bay in northern Quebec. This rock dates from 3.80 to 4.28 billion years old (see Figure 5.56). Issues were raised after discovery regarding the component samples used in the dating process, which are still not resolved. The locations of both deposits are shown in Figure 5.57.

Figure 5.55: *Acasta gneiss from Canada's Northwest Territories is 4.031 billion years old.*

Figure 5.56: Nuvvuagittuq *rocks found on the east coast of Canada's Hudson Bay may be 4.28 billion years old.*

Figure 5.57: This map shows the location of both finds: A = Acasta Gneiss, N = Nuvvuagittuq Greenstone Belt.

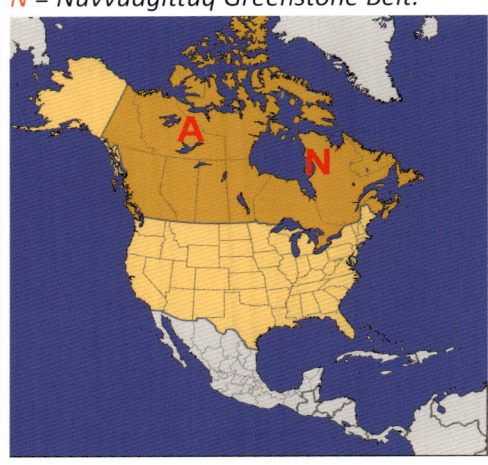

Rock Identification: Gneiss

A specimen can be identified as gneiss if it has the following characteristics.

Rock Identification Tips	Photographs or other identification information
Most of the mineral grains in gneiss are visible without magnification.	*Figure 5.58: The mineral grains in gneiss can be seen by the unaided eye.*
Gneiss is distinctive among other "banded" rocks because its minerals are not evenly distributed, and the ribbons of minerals have various widths.	*Figure 5.59: The mineral grains are not evenly distributed, which results in ribbons of minerals with different widths*

Rock Identification Tips	Photographs or other identification information
The metamorphic process caused minerals in gneiss to segregate into separate, typically light- and dark-colored layers.	*Figure 5.60:* Many gneiss specimens have alternating light and dark striations. 
Gneiss breaks into blocky chunks, rather than along the striated layers.	*Figure 5.61:* Gneiss breaks more like igneous rocks into blocky chunks rather than along the unorganized, foliated layers. 
Oftentimes the gneissic bands are folded and deformed, as a result of plate tectonic forces.	*Figure 5.62:* The bands in gneiss are often folded due to plate tectonic forces that deformed the layers. 

Rock Identification Tips	Photographs or other identification information
Gneiss is a tough hard rock with a hardness of between 6.0 and 7.0 on the Mohs Hardness Scale: it will scratch glass	*Figure 5.63:* Gneiss specimens will scratch glass. 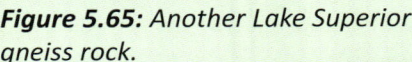
Gneiss does not contain fossils.	Verify the specimen does not contain fossils.
Gneiss is opaque.	Use a bright flashlight to verify the specimen is not translucent.

Gneiss Photographs

Figure 5.64: A gneiss specimen from a Lake Superior beach.

Figure 5.65: Another Lake Superior gneiss rock.

Figure 5.66: This gneiss cliff was significantly folded by plate tectonic forces.

Figure 5.67: This specimen has fewer platy minerals, so the foliation is less pronounced.

Figure 5.68: This gneiss specimen has foliation but no folding.

Figure 5.69: This specimen has dramatic foliated structure and folding.

Figure 5.70: This microscopic image shows how rounder quartz grains interbedded between flat grains (mica and hornblende).

Figure 5.71: This gneiss rock has fewer platy minerals, so it is less foliated.

Figure 5.72: Another gneiss from a Lake Superior beach.

Figure 5.73: Another beach-polished gneiss rock.

Figure 5.74: A gneiss specimen displaying the alternating light and dark layers.

Unakite

Unakite is a coarse-grained metamorphic rock that formed when granite (an igneous rock) was altered by hydrothermal activity (water under the Earth's surface heated by a magma chamber). During metamorphism, the black minerals in granite were changed and transformed into pistachio-green epidote. Pink potassium feldspar and clear to bluish-gray silica from the granite source rock are also present in unakite. This rock's interesting color blend and its hardness of between 6.0 and 7.0 on the Mohs Scale has helped it to become a popular lapidary material used to produce beads, jewelry, sculptures, tumbled stones, and other ornamental items (see Figure 5.75). The patterns, durability, and affordability of unakite have also made it a useful stone for architecture. Its most prominent use was trimming the front steps of the Smithsonian Museum of Natural History in Washington, D.C. (see Figure 5.76). Unakite was named in 1874 after the Unaka Range of mountains of North Carolina where it was first discovered.

Figure 5.75: A shaped and polished unakite heart.

In most cases, unakite formed in areas where continental plates collided. Fractures in the intrusive granite provided a conduit for the hot hydrothermal fluids to access the source rock. Under the right conditions, these hot fluids along with the temperatures and pressures of plate tectonic forces, caused the granite to metamorphose into unakite.

There can be confusion when trying to identify unakite. Some rocks contain green epidote mixed with red jasper. Although these specimens look similar, they are not unakite. Similar looking rocks in which no quartz is present are sometimes described as being unakite — but from a geological perspective — they are not the same stone.

Figure 5.76: Unakite was used to trim the steps at the Smithsonian Museum of Natural History in Washington D.C.

Rock Identification: Unakite

A specimen can be identified as unakite if it has the following characteristics.

Rock Identification Tips	Photographs or other identification information
The colors in unakite are mottled and blended and include pistachio-green, pink feldspar, and clear to bluish-gray quartz.	*Figure 5.77:* Unakite has green epidote, pink feldspar, and gray quartz.

Rock Identification Tips	Photographs or other identification information
Make sure the specimen has pink feldspar rather than red jasper.	**Figure 5.78:** Unakite has pink feldspar as one of its components (left image). **Figure 5.79:** Unakite does not contain red jasper (right image).  
Unakite is very compact and hard. Although there are obvious minerals, they are blended and do not show distinct individual crystal grains such as those that can be seen in granite, schist, or gneiss.	**Figure 5.80:** The minerals in unakite are blended and do not exhibit individual crystal grains, as can be seen in this polished specimen. 
Unakite rocks can have fractures and other erosional damage.	**Figure 5.81:** Unakite rocks can have fractures. 
Unakite has a hardness between 6.0 and 7.0 on the Mohs Hardness Scale, so it will scratch glass.	**Figure 5.82:** Unakite specimens will scratch glass. 
Unakite does not contain fossils.	Examine the specimen to verify it does not contain fossils.
Unakite is opaque.	Use a bright flashlight to verify the specimen is opaque.

Unakite Photographs

Please note: *The unakite specimens found on Great Lakes beaches have a limited supply because they were transported by glaciers from their source in Ontario, Canada between 8,000 and 10,000 years ago. Please limit what you collect and leave specimens for others to find and enjoy.*

Figure 5.83: This unakite specimen was found on a Lake Superior beach.

Figure 5.84: Another unakite from a Lake Superior beach.

Figure 5.85: This rough unakite specimen was found inland.

Figure 5.86: This unakite image was taken on a Lake Superior beach.

Figure 5.87: This unakite specimen has mostly epidote, less pink feldspar, and very little quartz.

Figure 5.88: A polished unakite stone.

Figure 5.89: A unakite rock that has suffered erosional and weathering damage from the waves and ice of Lake Superior.

Figure 5.90: A collection of unakite specimens.

Appendices

Appendix A: Instructions for Testing the Hardness of Rocks and Minerals

It is difficult to correctly identify rocks without significant field experience or sophisticated equipment. One method to obtain information useful in identifying a specimen is to use the Mohs Hardness Scale. This method was developed in 1822 by Frederich Mohs. The scale (see Figure A.1) is a chart of relative hardness of various minerals (1 = softest; 10 = hardest). This hardness test is easy and useful to perform as an aid in identifying a specimen.

What the hardness test involves is determining a specimen's resistance to being scratched. The test involves rubbing a mineral or item of known hardness against the specimen being tested. Because most people do not have the minerals listed in Figure A.1, some common items found around the house can also be used (see Figure A.2).

Figure A.1: Mineral hardness scale.

Mohs' Hardness Scale

1 = Talc
2 = Gypsum
3 = Calcite
4 = Fluorite
5 = Apatite
6 = Orthoclase Feldspar
7 = Quartz
8 = Topaz
9 = Corundum
10 = Diamond

Figure A.2: Mohs hardness of common household objects.

Mohs Hardness of Common Objects

Object	Hardness
Fingernail	2.0 - 2.5
Copper	3
Nail	4
Glass	5.5
Knife blade	5.0 - 6.5
Quartz	7

There are a couple of important things to point out about the hardness scale. Firstly, the scale in Figure A.1 is not mathematically linear. For example, diamond is much harder than corundum while calcite is only slightly harder than gypsum — even though in both cases, the minerals are one level of hardness different from each other on the scale. Also, if you use a penny to represent a hardness of 3.0, use a penny minted before 1983 when they were still made of mostly copper. Pennies minted today have only 2.5 percent copper and the rest is zinc.

The possible outcomes for performing a hardness test using only minerals are:
- ➤ If Specimen A scratches Specimen B, then Specimen A is harder than Specimen B.
- ➤ If Specimen A does not scratch Specimen B, then Specimen B is harder than Specimen A.
- ➤ If the two specimens are equal in hardness then small scratches might be produced on both specimens, or it might be difficult to determine if any scratches were produced.

To perform a hardness test, begin by locating a smooth, unscratched surface of the specimen. Use one hand to firmly hold the specimen being tested against a hard surface. If necessary, protect the surface (like a table top) against which you are pressing the specimen to protect it from being scratched. Firmly press a point of the object or mineral of known hardness against the specimen being tested. With firm pressure, drag the object across the surface of the specimen being tested. Carefully brush away any powder and examine the tested area. A scratch will be a distinct groove cut in the specimen's surface, not a mark on the surface that wipes away. Experience and practice will improve your testing skills and give you confidence. Some other suggestions are:

- Press hard enough to conduct the test, but not too hard as to break the specimen.
- The closer in hardness the two objects are, the more difficult it will be to produce a scratch.
- More force is required to test harder rocks than what is needed to test softer rocks.
- When possible, test a fresh surface of the rock because weathered surfaces may have been altered.
- Some minerals have different hardnesses, depending on if you are testing with the grain or against it.
- The hardness of rocks can vary, so even specimens of the same rock type may have a slightly different hardness depending on the exact mineral compositions and the section of the rock being tested.
- Testing the hardness of sedimentary rocks can be challenging because the natural cement fusing the components may be softer than the components. Sedimentary rocks can also break apart during testing.
- The hardness of fine-grained rocks may equal the average hardness of the rock's mineral components.
- Use a hand lens to get a better look at the testing site.
- Repeat the test a second time to verify your results.

Note: Thanks to Dr. William Cordua for some of this information.

Appendix B: Rock Cycle Diagram

A rock is a solid object made up of different minerals. Rocks are generally not uniform in the organization of its mineral grains nor in the percentage of these mineral components. Because rocks of the same variety can differ, rocks cannot be described by scientific formulas. Scientists instead classify rocks by how they formed. There are three major types of rocks: igneous, sedimentary and metamorphic.

The rock cycle is a concept used to explain how these three basic rock types are related and how geologic processes over time cause them to form. Plate tectonic activity, weathering, and erosional processes are responsible for the continued formation and recycling of rocks.

Figure A.3: Rock cycle diagram.

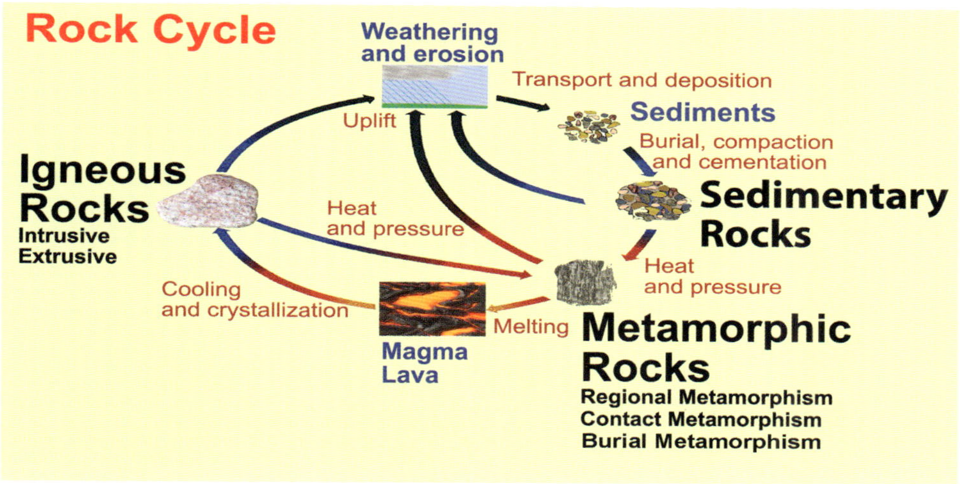

Appendix C: Quick Reference Identification Chart

- ▢ Intrusive Igneous Rocks, pages 15-59
- ▢ Extrusive Igneous Rocks, pages 16-92
- ▢ Amygdaloidal Minerals, pages 93 to 119
- ▢ Sedimentary Rocks, pages 120-184
- ▢ Metamorphic Rocks, pages 185-212

■ = Mostly true
S = Sometimes true

Characteristic	Granite	Pegmatite	Granodiorite	Diorite	Gabbro	Syenite	Basalt	Vesicular Basalt
Mineral grain size and arrangement								
The mineral grains or crystals are visible to the unaided eye.	■	■	■	■	■	■		
The mineral grains are not visible without magnification.							■	■
The mineral grains are arranged in a random pattern.	■	S	■	■	■	■		
The mineral grains are not evenly distributed and vary in size.							S	S
The specimen's components have a horizontal arrangement.								
Translucency								
The specimen is opaque.	■	■	■	■	■	■	■	■
The specimen is translucent.								
Color								
The specimen's color is more dark than light.					■	S	■	■
The specimen's color is more light than dark.	■	■	■	■		■		
The specimen is monotone in color.							■	■
The specimen has components with different colors.	■	■	■	■	■	■		
Pockets								
The specimen has open unfilled holes/pockets.							S	■
The specimen has holes/pockets that are filled with minerals.							S	
The specimen does not have any holes/pockets.	■	■	■	■	■	■	S	
Surface Characteristics								
The specimen's surface has a waxy luster.								
Visible flat crystal faces are shiny when exposed to light.	S	S	S	S	S	S		
The specimen has conchoidal fractures.								
The specimen is smooth to the touch.	S	S	S	S	S	S	■	
The specimen fizzes when subjected to drops of acid.								
The specimen fluoresces when exposed to UV light.	S	S				S		
The specimen feels like fine-grit sandpaper.								
The specimen can easily be split into horizontal slabs.								
Hardness Test								
The specimen is hard and will scratch glass.	■	■	■	■	■	■	■	
The specimen can be scratched with a nail.								
The specimen can be scratched with a knife blade.								
Fossils								
The specimen does not contain fossils.	■	■	■	■	■	■	■	
The specimen does contain fossils.								

Note: This chart is a quick reference to provide general characteristics for the rocks and minerals included in this book. The characteristics may not always apply.

Appendix D: Geologic Timeline

Although rocks and minerals have been mined on a small scale for thousands of years, large commercial mining operations did not begin until the 1600s. Miners recognized certain tendencies in the way rocks were organized such as: sedimentary rocks were deposited in horizontal layers; younger sedimentary layers were deposited on top of older layers. In the early 1900s, the first geologic time scale that included absolute dates was published by the British geologist Arthur Holmes. The timeline described the timing and relationships of events in geologic history. It was developed through the study of physical rock layers and their relationships as well as the times when different organisms appeared, evolved, and became extinct through the study of fossilized remains and imprints. Figure A.4 shows the major hierarchical chunks of time, called eons. Figure A.5 is included as an example to show some of the key geologic events that occured in the Great Lakes region.

Figure A.4: The eons of the Earth's geologic timeline are shown in this diagram.

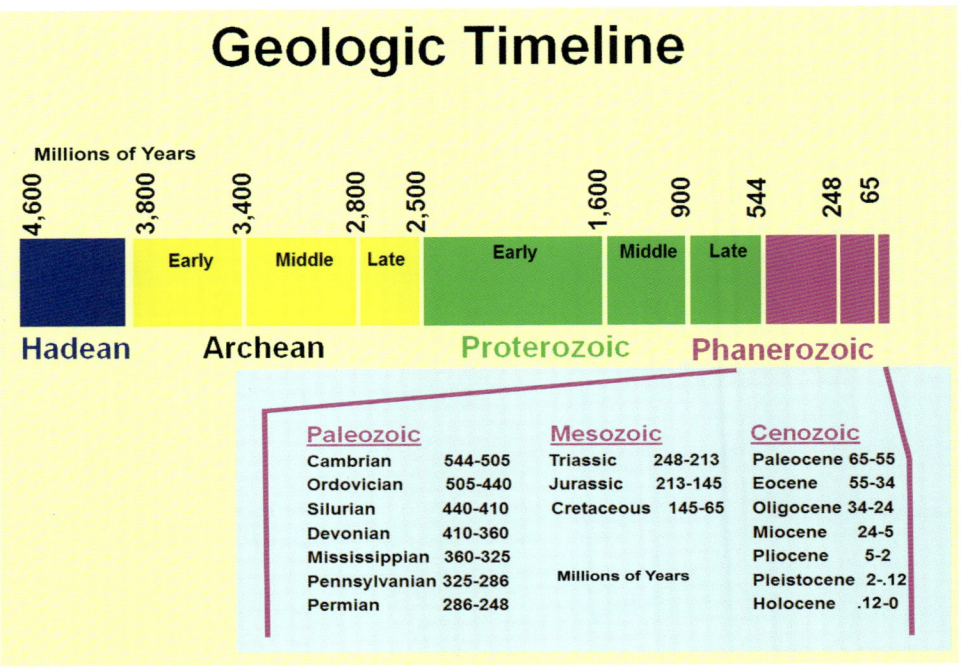

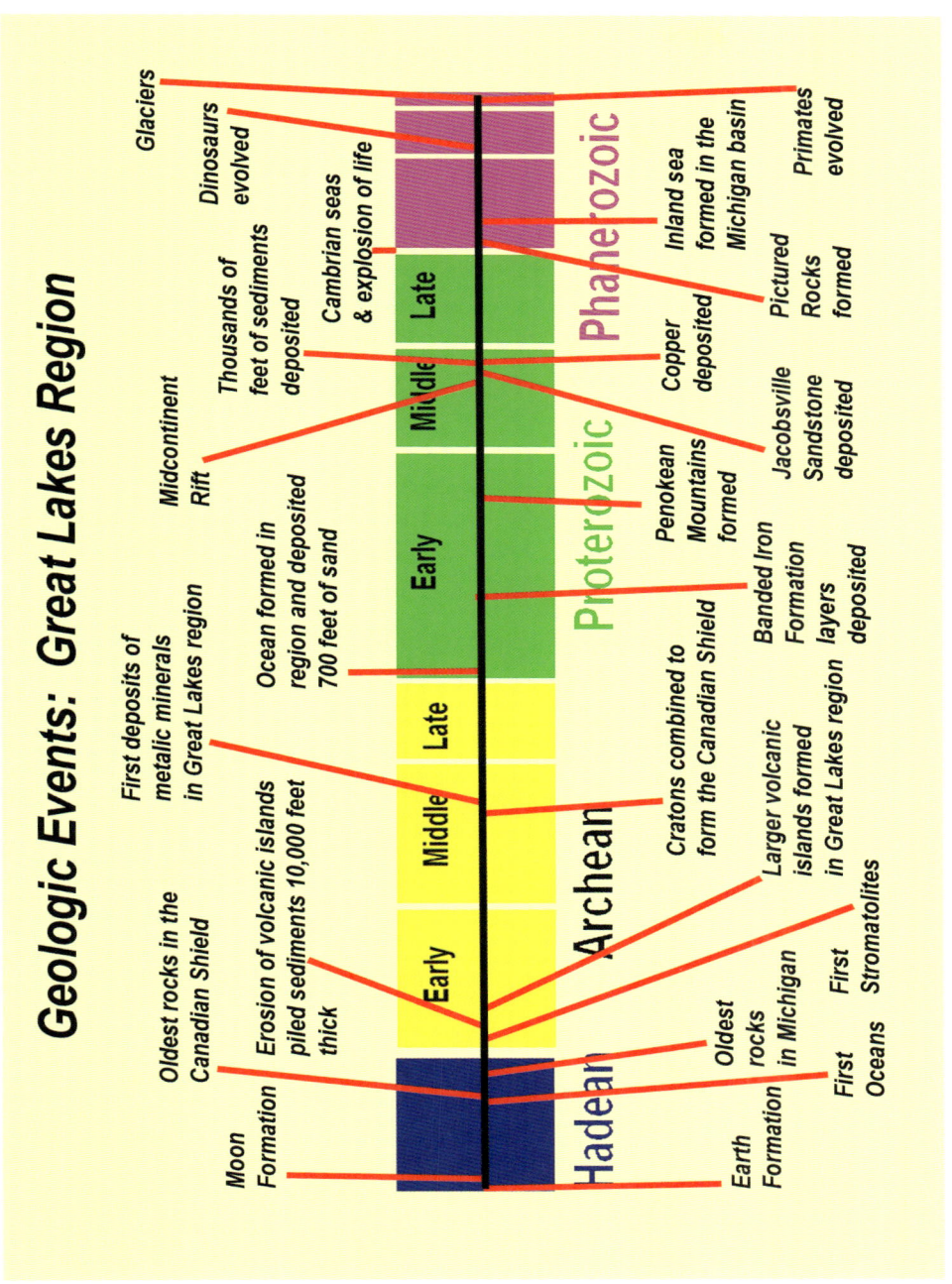

Figure A.5: This chart shows the significant geologic events in the history of the Great Lakes region. It is included as an example.

Appendix E: Bibliography

A considerable amount of research was done for this book. There are hundreds of additional books and articles that could have been reviewed, but a decision was made to stop the research phase of this project when it appeared I had enough information in my data base. All the sources consulted during this initial research phase are listed below. In addition, during the writing phase, Internet searches were done to seek additional information about specific rocks or to improve my own understanding about geologic history. This was a free-flow investigative process meant to improve verify the accuracy of the book's information. These sources were not recorded.

Books, Articles, and Pamphlets

1. Bornhorst, Theodore J. and William I. Rose (1994 to 1999). *Self-guided geological field trip to the Keweenaw Peninsula, Michigan.* Institute on Lake Superior Geology, Houghton, MI.
2. Boyer, Robert E. and William H. Mathews III (1984). *Wards GEO-logic System: Metamorphic Rocks, Sedimentary Rocks, and Igneous Rocks.*
3. Conway, W. Rick, (1957),.*Origins of Rocks and Minerals.* Earth Exploration Company, New Albany, IN.
4. Dake, H.C., Frank L. Fleener, and Ben Hur Wilson (1938). *Quartz Family Minerals: A Handbook for the Mineral Collector.* McGraw-Hill Book Company, Inc., New York, NY.
5. Dana, Edward S. and Cornelius S. Hurlbut, Jr. (1949). *Minerals and How to Study Them*. John Wiley & Sons, Inc., New York, NY.
6. Dietrich, R.V. (1980). *Stones: Their Collection, Identification, and Uses.* W.H. Freeman and Company.
7. Fenton, Carroll L. and Mildred A. Fenton (1940, 1968, 2003). *The Rock Book.* Dover Publications, Inc., Mineola, NY.
8. Gauthier, Kevin and Bruce Mueller (2007). *Lake Superior Rock Picker's Guide.* The University of Michigan Press, Ann Arbor, MI.
9. Halda, S.K. *(2017). Geology and Geochemistry, Platinum-Nickel-Chromium Deposits.*
10. Haldar, S.K. and Josip Tišljar (2014). *Igneous Rocks, Introduction to Mineralogy and Petrology.*
11. Heinrich, W. Wm. (1976). *The Mineralogy of Michigan*. Michigan Geological Survey Division, Lansing, MI.
12. Honovich, Nancy, *Rocks & Minerals.* National Geographic Kids Ultimate Explorer Field Guide.
13. KumarShukla, Matsyendra and AnupamSharma (June 2018). *A brief review on breccia: its contrasting origin and diagnostic signatures.* Solid Earth Sciences, Volume 3, Issue 2, Pages 50-59.
14. LaBerge, Gene (1994). *Geology of the Lake Superior Region.* Geoscience Press, Inc., Tucson.
15. Lawton, Rebecca, Diana Lawton, and Susan Panttaja (1997). *Discover Nature in the Rocks: Things to know and things to do.* Stackpole Books.
16. Michigan Geologic Survey (1964). *Our Rock Riches: A Selected Collection of Reprinted Articles on Michigan's Mineral Resources By Various Authors.* Lansing, MI.

17. Pearl, Richard M. (1967). *Gems, Minerals, Crystals, and Ores.* Golden Press, New York, NY.
18. Pellant, Chris (1992). *Rocks and Minerals: The visual guide to more than 500 rocks and minerals from around the world.* A DK Publishing Book.
19. Pough, Frederick H. (1953). *A Field Guide to Rocks and Minerals*. Houghton Mifflin Company, Boston, MA.
20. Rose, William, Erika Vye, and Valerie Martin (2017). *How the Rock Connects Us: A Geoheritage Guide to Michigan's Keweenaw Peninsula and Isle Royale.* Houghton, MI, Isle Royale & Keweenaw Parks Association.
21. Stensaas, Mark S. (2000). *Rock Picker's Guide to Lake Superior's North Shore.*
22. White, Anne Terry (1955). *All About Rocks and Minerals*. Random House, New York, NY.
23. Zim, Herbert S. and Paul R. Shaffer (1957). *Rocks and Minerals: A Guide to Familiar Minerals, Gems, Ores and Rocks.* Golden Press, New York, NY.
24. Zim, Herbert S., Paul R. Shaffer, Raymond Perlman (1957). *Rocks and Minerals: A Guide to Minerals, Gems, and Rocks*. Golden Press, Inc., New York, NY.

Internet Sources
1. Alden, Andrew, https://www.thoughtco.com/what-is-chert-1441025.
2. Earth Science Teachers Association. Basalt Rocks, https://www.windows2universe.org/earth/geology/ig_basalt.html.
3. Fitton, Godfrey (2020). Basalt and Related Rocks. https://www.sciencedirect.com/topics/earth-and-planetary-sciences/basalt.
4. Gemselect, https://www.gemselect.com/english/gem-info/carnelian/carnelian-info.php.
5. Geology Science (13th November 2020). Breccia. https://geologyscience.com/rocks/breccia/.
6. Geologyscience.com, https://geologyscience.com/rocks/sedimentary-rocks/chert/.
7. John, James St.: information was used from over 100 of his Wikipedia posts.
8. Julig, P.J., L.A.Pavlish, C. Clark and R.G.V. Hancock, Chemical Characterization and Sourcing of UpperGreat Lakes Cherts by INAA, https://www.ontarioarchaeology.org/Resources/Publications/oa54-3-julig.pdf.
9. King, Hobart. What is Breccia, How Does it Form, and What is its Composition? https://geology.com/rocks/breccia.shtml.
10. King, Hobart M. What Is Basalt, How Does It Form, and How Is It Used? https://geology.com/rocks/basalt.shtml.
11. King, Hobart M., https://geology.com/rocks/breccia.shtml.
12. King, Hobart M., https://geology.com/rocks/chert.shtml.
13. Learning Geology. http://geologylearn.blogspot.com/2015/03/basalt.html.
14. Mineralogy Society of America, http://www.minsocam.org/msa/collectors_corner/vft/mi2b.htm.
15. National Earth Science Teachers Association, https://www.windows2universe.org/earth/geology/ig_basalt.html.
16. National Geographic, https://www.nationalgeographic.org/article/magma-role-rock-cycle/.
17. National Geographic, https://www.nationalgeographic.org/encyclopedia/magma/.
18. OhRanger.com, http://www.ohranger.com/pictured-rocks/geology.
19. Rose, William, http://www.geo.mtu.edu/KeweenawGeoheritage.

20. OpenGeology.org, https://opengeology.org/Mineralogy/4-crystals-and-crystallization/.
21. Sandatlas, https://www.sandatlas.org/basalt/.
22. Sandatlas.org, https://www.sandatlas.org/chert/.
23. Sasso, Andrew (2016), *Geochemical and Petrological Characterizations of Peridotite and Related Rocks in Marquette County, Michigan*. Master's Theses. https://scholarworks.wmich.edu/masters_theses/687.
24. Schaetzl, Randall, http://geo.msu.edu/extra/geogmich/rift_zone.html.
25. Stein, Seth, et al (4 August 2016). https://eos.org/features/new-insights-into-north-americas-midcontinent-rift.
26. Stoffer, Phil, https://geologycafe.com/.
27. StoneContract, http://stonecontract.eu/wiki/about-gem-stone/about-carnelian/.
28. The University of Auckland, New Zealand, https://flexiblelearning.auckland.ac.nz/rocks_minerals/rocks/basalt.html.
29. The University of Auckland, New Zealand, https://flexiblelearning.auckland.acWnz/rocks minerals/rocks/breccia.html.
30. Thoughtco.com, https://www.thoughtco.com/pictures-of-chert-4122739.
31. Wikipedia, https://en.wikipedia.org/wiki/Basalt.
32. Wikipedia, https://en.wikipedia.org/wiki/Carnelian.
33. Wikipedia, https://en.wikipedia.org/wiki/Chert.
34. Wikipedia, https://en.wikipedia.org/wiki/Lava.
35. Wikipedia, https://en.wikipedia.org/wiki/Magma.
36. Wikipedia, https://en.wikipedia.org/wiki/Mantle_plume.
37. Wikipedia, https://en.wikipedia.org/wiki/Midcontinent_Rift_System.
38. Wikipedia, https://en.wikipedia.org/wiki/Breccia.

Appendix F: Figure Attributions

Due to the large number of images included in this book, the attributions listed below do not include the photographs and diagrams I have created, but instead only include those acquired from other sources. Thus, if there is a figure number not listed, credit can be attributed to Karen Brzys. This decision saves several pages of printing. To also save paper, this list is in small type. If you want the whole list (including cites for my work), or if you would like this list in a larger font size, please send an email to Karen@agatelady.com.

Introduction
5. Sander van der We, https://commons.wikimedia.org/wiki/File:(61-365)_Can_you_imagine%3F_(5320329773).jpg.
6. Dr. K.L. Milliken, University of Texas at Austin. Used with permission.
9. Dcrjsr, https://commons.wikimedia.org/wiki/File:Fossil_rock_Rocky_Ridge_Las_Trampas.jpg.

Chapter 1: Source of Minerals
1. NASA image, https://commons.wikimedia.org/wiki/File:Supernova_Remnant_W49B_in_x-ray,_radio,_and_infrared.jpg.
2. NASA drawing, https://commons.wikimedia.org/wiki/File:Brown_dwarf_OTS_44_with_disc.jpg.
3. Shutterstock_92871580.
4. John W. Valley, University of Wisconsin, Madison. Image used with permission.
5. Shutterstock_97202195.
6. NASA image, https://commons.wikimedia.org/wiki/File:Pasteur_D_crater_Apollo_15.jpg.
7. Shutterstock_116489824.
8. Karen Brzys diagram with information from Citronade, https://commons.wikimedia.org/wiki/File:Moon_-_Giant_Impact_Hypothesis_-_Simple_model.png.
9. Shutterstock_90878378.
10. Willem van Aken, CSIRO, https://commons.wikimedia.org/wiki/File:CSIRO_ScienceImage_4203_A_bluegreen_algae_species_Cylindrospermum_sp_under_magnification.jpg.

Chapter 2: Intrusive Igneous Rocks
3. Distress.bark, https://commons.wikimedia.org/wiki/File:Mount_Gould_from_Grinnell_Glacier_Trail.JPG.
4. Jonathan.s.kt, https://commons.wikimedia.org/wiki/File:Geological_Dike_Cross-Island_Trail_Alaska.jpg.
5. James St. John, https://commons.wikimedia.org/wiki/File:Merced_River_(Yosemite_Valley,_Sierra_Nevada_Mountains,_California,_USA)_3_(19847384318).jpg.
6. Karen Brzys image with information from https://opengeology.org/Mineralogy/4-crystals-and-crystallization/.
7. Roll-Stone, https://commons.wikimedia.org/wiki/File:Rosa-Beta-Granit.jpg.
8. Thomas Bresson, https://sk.m.wikipedia.org/wiki/S%C3%BAbor:Thomas_Bresson_-_Granite_vu_au_microscope_(by).jpg.
10. James St. John, https://commons.wikimedia.org/wiki/File:Half_Dome_(Sierra_Nevada_Mountains,_California,_USA)_26.jpg.
11. Dean Franklin, https://commons.wikimedia.org/wiki/File:Dean_Franklin_-_06.04.03_Mount_Rushmore_Monument_(by-sa).jpg.
12. Shutterstock_114730579.
14. Karen Brzys diagram with information from https://kaiserscience.wordpress.com/earth-science/earths-layered-structure/mantle-convection/.
15. Shutterstock_609415880.
16. James St. John, https://commons.wikimedia.org/wiki/File:Potassium_feldspar_(32499528651).jpg.
17. James St. John, https://commons.wikimedia.org/wiki/File:Biotite_mica_2_(31739438210).jpg.
18. James St. John, https://commons.wikimedia.org/wiki/File:Hornblende_amphibole_(31733452843).jpg.
20. Dominicus Johannes Bergsma, https://commons.wikimedia.org/wiki/File:Gletsjerpad_naar_de_Morteratschgletsjer_12-09-2019._(d.j.b)_28.jpg.
21. USGS, https://commons.wikimedia.org/wiki/File:Kaolinite_-_USGS_bws00008.jpg.
24. James St. John, https://commons.wikimedia.org/wiki/File:Granite_9_(48674142571).jpg.
27. James St. John, https://commons.wikimedia.org/wiki/File:%22Columbia_Pink_Granite%22_(porphyritic_granite,_Conway_Granite,_Jurassic;_east_of_Tinkerville,_New_Hampshire,_USA)_(14618798910).jpg.

32. James St. John, https://commons.wikimedia.org/wiki/File:Granite_41_(49200495993).jpg.
33. James St. John, File:Granite (Silver Plume Granite, Mesoproterozoic, 1.42 Ga; Rt. 36 roadcut, Larimer County, Colorado, USA) 1(31567132672).jpg - Wikimedia Commons.
36. David Monniaux, https://commons.wikimedia.org/wiki/File:Granite_Yosemite_P1160483.jpg.
37. Photograph recycled from a previous project. Source unknown.
38. James St. John, https://commons.wikimedia.org/wiki/File:Nara_Brown_Granite_(charnockite)_Quebec.jpg.
39. S Rae, https://commons.wikimedia.org/wiki/File:Pink_Granite_-_Flickr_-_S._Rae.jpg.
40. James St. John,https://commons.wikimedia.org/wiki/File:%22Rockville_White_Granite%22_(porphyritic_granite,_Rockville_Granite,_late_Paleoproterozoic,_1.73_to_1.78_Ga;_quarry_near_Rockville,_Minnesota,_USA)_1_(25902855894).jpg.
41. Shutterstock 1040188849.
44. James St. John, File:Emeralds in pegmatitic granite 8 (37992559234).jpg - Wikimedia Commons.
45. James St. John, https://commons.wikimedia.org/wiki/File:Garnets_and_tourmaline_in_pegmatitic_granite_(Crabtree_Pegmatite,_Devonian;_Crabtree_Mountain,_Mitchell_County,_North_Carolina,_USA)_1_(25144467358).jpg.
46. Jonas Börje Lundin, https://commons.wikimedia.org/wiki/File:Hellglimmer_in_Pegmatit.jpg.
47. William Cordua, photograph used with permission.
51. James St. John, https://commons.wikimedia.org/wiki/File:Pegmatitic_granite_(Precambrian;_Alexandra_Mine,_Broken_Hill_Block,_New_South_Wales,_Australia)_2_(26487685125).jpg.
52. James St. John, https://commons.wikimedia.org/wiki/File:Graphic_granite_(runite)_(Eight_Mile_Park_Pegmatite_District,_Mesoproterozoic,_1.430_to_1.474_Ga;_roadcut_along_3A_Road,_north_of_Royal_Gorge_Bridge,_west_of_Canon_City,_south-Central_Colorado,_USA)_10_(48751697192).jpg.
53. Géry Parent, https://commons.wikimedia.org/wiki/File:Mica,_feldspar.jpg.
54. William Cordua, photograph used with permission.
56. William Cordua, photograph used with permission.
58. Photograph recycled from a previous project. Source unknown.
60. James St. John, File:Garnets and tourmaline in pegmatitic granite (Crabtree Pegmatite, Devonian; Crabtree Mountain, Mitchell County, NorthCarolina, USA) 2 (27237621539).jpg - Wikimedia Commons.
61. James St. John, https://commons.wikimedia.org/wiki/File:Orbicular_granodiorite_(Neoarchean,_2.687_Ga;_Warroan_Hill,_West_Australia)_5_(48751302333).jpg.
62. James St. John, https://commons.wikimedia.org/wiki/File:Orbicular_granodiorite_(Neoarchean,_2.687_Ga;_Warroan_Hill,_West_Australia)_4_(48751624851).jpg.
63. James St. John, https://commons.wikimedia.org/wiki/File:Granodiorite_(Giant_Forest_Granodiorite,_mid-Cretaceous,_97-102_Ma;_Moro_Rock,_Sequoia_National_Park,_California,_USA)_8_(16611992558).jpg.
64. James St. John, https://commons.wikimedia.org/wiki/File:Granodiorite_(Giant_Forest_Granodiorite,_mid-Cretaceous,_97-102_Ma;_Moro_Rock,_Sequoia_National_Park,_California,_USA)_7_(16592281297).jpg.
65. Nicolas Eynaud, https://commons.wikimedia.org/wiki/File:Granodiorita_-_Grossiment_x40_-_Lutz_polarizada.png.
68. Modified from: James St. John, https://commons.wikimedia.org/wiki/File:Xenolith_in_andesite_(Tertiary;_Yellowstone,_Wyoming,_USA)_14_(48555386541).jpg.
69. Metropolitan Museum of Art, File:Kohl Jar MET 26.8.38ab.front.jpg - Wikimedia Commons.
70. Michael C. Rygel, https://commons.wikimedia.org/wiki/File:Diorite2.tif.
71. Recycled from a previous project. Source unknown.
72. Shutterstock_786743608.
74. Slim Sepp, https://commons.wikimedia.org/wiki/File:Diorite_pmg_ss_2006.jpg.
77. Recycled from a previous project. Source unknown.
79. Shutterstock_238889983.
84. Recycled from a previous project. Source unknown.
85. Amcyrus2012, https://commons.wikimedia.org/wiki/File:Diorite_MA.JPG.
86. Michael C. Rygel, File:Diorite3.tif - Wikimedia Commons.
87. Michael C. Rygel, File:Diorite1.tif - Wikimedia Commons.
88. Daniele.51, File: Diorite orbicolare.jpg—Wikimedia Commons.
89. Pekachu, Orbicular_diorite_from_Ainoura,_Taku_at_History_and_Folklore_Museum.jpg.
90. Euphra, File: N_Mt_Schimansky_sheared_diorite_2.jpg – Wikimedia Commons.
91. Euphra, File: wikipedia/commons/0/0d/Janke_Nunatak_diorite_layering_portrait.jpg.
92. Randolph Black, wikipedia/commons/6/64/Diorite_01_10x_%2827879456709%29.jpg.
94. James St. John, https://commons.wikimedia.org/wiki/File:Spheroidally_weathered_diabase_(Haverstraw,_southeastern_New_York_State,_USA).jpg.
95. Karen Brzys diagram with information from https://geology.com/rocks/gabbro.shtml.

97. Julien Leuthold, https://commons.wikimedia.org/wiki/File:Gabbro_from_Rum_in_Scotland_-_Thin_Section.jpg.
107. James St. John, https://commons.wikimedia.org/wiki/File:Olivine_gabbro_(Pigeon_Point_Sill,_Mesoproterozoic,_~1.1_Ga;_Pigeon_Point,_Minnesota,_USA)_(40770721514).jpg.
112. This wiki image was recycled from a previous project. Source unknown.
113. Max.kit, https://commons.wikimedia.org/wiki/File:Gabbro_cap_zuid_afrika.jpg.
115. Mark A. Wilson, https://commons.wikimedia.org/wiki/File:GabbroRockCreek1.jpg.
116. Anders Damber, https://commons.wikimedia.org/wiki/File:Gabbro_Norrbotten.jpg.
117. Michael C. Rygel via Wikimedia Commons, https://commons.wikimedia.org/wiki/File:Melagabbro1.tif
118. Shutterstock 139151960.
119. Shutterstock_770764387.
120. Wilson44691, File:GabbroRockCreek2.jpg - Wikimedia Commons.
121. Euphro, File:Mount Eissenger stoped gabbro (o'expd).jpg - Wikimedia Commons.
122. James St. John, https://commons.wikimedia.org/wiki/File:Layered_Gabbro_of_the_Duluth_Complex_in_Minnesota_USA_01.jpg.
124. Philippe Giabbanelli, https://commons.wikimedia.org/wiki/File:Sodalite_(Mineral).jpg.
125. Parent Géry, https://commons.wikimedia.org/wiki/File:Hackmanite,_winchite_sous_UVL_2.JPG.
128. Natasha Sel, https://commons.wikimedia.org/wiki/File:Amphibole_in_syenites.jpg.
134. James St. John, https://commons.wikimedia.org/wiki/File:Nepheline_syenite_(questionably_from_Wausau,_Wisconsin,_USA).jpg.
136. Jstuby, https://commons.wikimedia.org/wiki/File:Nepheline_syenite_NJ.jpg.
137. James St. John, , https://commons.wikimedia.org/wiki/File:Syenite_(Cuttingsville_Complex,_mid-Cretaceous,_97_Ma;_southwest_of_Cuttingsville,_southern_Vermont,_USA)_(16563008099).jpg.

Chapter 3: Extrusive Igneous Rocks
3. Karen Brzys diagram modified from https://commons.wikimedia.org/wiki/File:North_america_basement_rocks.png.
4. Karen Brzys diagram modified from John Goodge, https://commons.wikimedia.org/wiki/File:Rodinia_reconstruction.jpg.
10. USGS, https://prd-wret.s3.us-west-2.amazonaws.com/assets/palladium/production/s3fs-public/thumbnails/image/multimediaFile-2353.jpg.
13. James St. John. https://commons.wikimedia.org/wiki/File:Labradorite_(Wiborg_Batholith,_1633_Ma;_Kymi_Province,_Finland)_7.jpg.
14. Jan Helebrant, https://commons.wikimedia.org/wiki/File:Pyroxene_(26043678351).jpg.
15. James St. John, https://commons.wikimedia.org/wiki/File:Hydrothermal_quartz_crystals_(Late_Pennsylvanian_to_Permian;_Coleman_Quartz_Mine,_Arkansas,_USA)_3_(30421915874).jpg.
16. James St. John, https://commons.wikimedia.org/wiki/File:Olivine_(31834223564).jpg.
17. James St. John, https://commons.wikimedia.org/wiki/File:Hornblende_amphibole_(31733452843).jpg.
18. NASA: https://www.jpl.nasa.gov/spaceimages/details.php?id=PIA02982.
19. Chmee2, https://commons.wikimedia.org/wiki/File:Basalt-microscop-crossed_polars.jpg.
20. Uncle.Capung, https, ://commons.wikimedia.org/wiki/File:Basalt_Plagioclase_and_Pyroxene.jpg.
22 James St. John, https://commons.wikimedia.org/wiki/File:Baptism_River_Basalt_(North_Shore_Volcanic_Group,_Mesoproterozoic,_~1097_Ma;_Split_Rock_Lighthouse,_Minnesota,_USA)_2_(22633565687).jpg.
23. https://prd-wret.s3.us-west-2.amazonaws.com/assets/palladium/production/s3fs-public/vhp_img2858.jpg.
24. Ekaterina Vasyagina, https://wikimediafoundation.org/news/2018/12/17/lose-yourself-in-our-planets-beauty-with-the-winners-of-wiki-loves-earth/.
25. James St. John, https://commons.wikimedia.org/w/index.php?search=vesicular+basalt+fillmore+craters%2C+holocene&title=Special%3ASearch&go=Go&ns0=1&ns6=1&ns12=1&ns14=1&ns100=1&ns106=1#/media/File:Vesicular_basalt_(Fillmore_Craters,_Holocene;_Black_Rock_Desert,_due_west_of_Fillmore,_Sevier_Desert,_west-central_Utah,_USA)_1_(15036417195).jpg.
26. Shutterstock_424005253.
27. Digon3, https://commons.wikimedia.org/wiki/File:Scoria_edit.jpg.
28. Peka, https://commons.wikimedia.org/wiki/File:Scoria_on_a_road_in_Michigoshi,_Tara_01.jpg.
29. Jon Zander, https://commons.wikimedia.org/wiki/File:Scoria_Macro.JPG.
30. Randolph Black, https://commons.wikimedia.org/wiki/File:Scoria_01_10x_(39669390031).jpg.
32. James St. John, https://commons.wikimedia.org/wiki/File:Porphyritic_basalt_(Middle_Pleistocene,_670_ka;_Black_Rock_Volcano,_Millard_County,_Utah,_USA)_1_(30869888644).jpg.
33. James St. John, https://commons.wikimedia.org/wiki/File:Olivine_basalt_(Cedar_Canyon,_Iron_County,_Utah,_USA)_6_(48679986543).jpg.

34. Tom Shearer photograph from *Agates Inside Out* (rights co-owned by Karen Brzys).
39. Tom Shearer photograph from *Agates Inside Out* (rights co-owned by Karen Brzys).
40. James St. John, https://commons.wikimedia.org/wiki/File:Amygdaloidal_basalt_1.jpg.
42. Photograph from a previous project. Source unknown.
43. Shutterstock_255784546.
44. Shutterstock_130253333.
45. James St. John, https://commons.wikimedia.org/wiki/File:Olivine_basalt_(Cedar_Canyon,_Iron_County,_Utah,_USA)_7_(48680327561).jpg.
50. http://www.earth.ox.ac.uk/~oesis/micro/Igneous/index.html#img=basalt1-ppl_pm15-17.jpg.
51. http://www.earth.ox.ac.uk/~oesis/micro/sedimentary/index.html#img=greywacke_pm19-16.jpg.
52. James St. John, https://commons.wikimedia.org/wiki/File:Amygdaloidal_basalt_(Portage_Lake_Volcanic_Series,_upper_Mesoproterozoic,_1.093_to_1.097_Ga;_Keweenaw_Peninsula,_Upper_Peninsula_of_Michigan,_USA)_12.jpg.
53. James St. John, https://commons.wikimedia.org/wiki/File:Amygdaloidal_basalt_(upper_Portage_Lake_Volcanic_Series,_upper_Mesoproterozoic,_~1.094_Ga;_Delaware_Copper_Mine,_Upper_Peninsula_of_Michigan,_USA)_4.jpg.
55. James St. James, https://commons.wikimedia.org/wiki/File:Vesicular_basalt_(16128028613).jpg.
56. Shutterstock_58930969.
63. Rohitjahnavi, https://commons.wikimedia.org/wiki/File:Volcanic_andesite_rock_from_Narcondam_Island.jpg.
64. https://blogs.nvcc.edu/mineralogy/files/2020/10/50HD0276.jpg.
65. James St. John, https://commons.wikimedia.org/wiki/File:Andesite_(near_Belleville,_Mineral_County,_Nevada,_USA).jpg.
66. James St. John, File:Porphyritic andesite (Kate Peak Formation, Middle Miocene; Lyon County, western Nevada, USA)(15661069958).jpg - Wikimedia Commons.
67. NASA, Student Project: Describe Rocks Like a NASA Scientist | NASA/JPL Edu.
68. Marie Helliwell, https://commons.wikimedia.org/wiki/File:M%C4%81%C4%93,1933.379,19599.8,_(a).jpg.
69. Marek Novotňák, File:Weathered andesite.jpg - Wikimedia Commons.
70. James St. John, File:Xenolith in andesite (Tertiary; Yellowstone, Wyoming, USA) 14 (48555386541).jpg - Wikimedia Commons.
72. Tom Shearer photograph from *Agates Inside Out* (rights co-owned by Karen Brzys).
73. Siim Sepp, https://commons.wikimedia.org/wiki/File:Rhyolite_pmg_ss_2006.jpg.
74. Ji-Elle, File:La Salle-Les Fossottes-Ebauches de meules de moulins à bras en rhyolite (1).jpg - Wikimedia Commons.
75. Michael C. Rygel, https://commons.wikimedia.org/wiki/File:PinkRhyolite.tif.
79. Rockman 12, https://commons.wikimedia.org/wiki/File:Orbicular_Rhyolite.JPG.
81. Allandale_Rhyolite_Lyttelton_New_Zealand.jpg, JPR46, https://commons.wikimedia.org/wiki/File:Allandale_Rhyolite_Lyttelton_New_Zealand.jpg.
82. Michael C. Rygel, https://commons.wikimedia.org/wiki/File:Flow_banded_rhyolite.JPG.
83. Albertacce,_Corsica.jpg, Cardioceras, https://commons.wikimedia.org/wiki/File:Flow-banded_rhyolite,_Albertacce,_Corsica.jpg.
84. James St. John, https://commons.wikimedia.org/wiki/File:Garnet_in_rhyolite_(Garnet_Hill_Rhyolite,_Late_Eocene,_36-37_Ma;_Garnet_Hill,_White_Pine_County,_Nevada,_USA).jpg.
87. Amcyrus2012, https://commons.wikimedia.org/wiki/File:Pink_Rhyolite_Student_Sample.JPG.
88. James St. John, https://commons.wikimedia.org/wiki/File:Porphyritic_rhyolite_(Pleistocene;_Yellowstone,_Wyoming,_USA)_(26755758577).jpg.
92. Amcyrus2012, https://commons.wikimedia.org/wiki/File:Rhyolite_Colorado.JPG.
93. Shutterstock_436468300.
94. Shutterstock_391713463.
96. James St. John, https://commons.wikimedia.org/wiki/File:Columnar-jointed_rhyolitic_obsidian_lava_flow_(Roaring_Mountain_Member,_Plateau_Rhyolite,_Upper_Pleistocene,_~59_ka;_Obsidian_Cliff,_Yellowstone,_Wyoming,_USA)_3_(46092909084).jpg.
101. Amcyrus2012, https://commons.wikimedia.org/wiki/File:Pumice.JPG.
102. Norbert Nagel, https://commons.wikimedia.org/wiki/File:Lanzarote_-_stones_of_a_wall_-_pumice_stone.jpg.
103. Mauro Cateb, https://commons.wikimedia.org/wiki/File:Pumice_stones.JPG.
104. Jan Helebrant, https://commons.wikimedia.org/wiki/File:Chalcedony_SiO2_(24322909894).jpg.
105. Karen Brzys diagram with information from https://upload.wikimedia.org/wikipedia/commons/d/d8/Lithic_flake.png.

107. Shutterstock_1028083981.
108. Shutterstock_192824450.
109. Tom Shearer photograph from *Agates Inside Out* (rights co-owned by Karen Brzys).
110. James St. John, https://commons.wikimedia.org/wiki/File:Chalcedony-lined_geode_(Atlas_Mountains,_Morocco)_(34485358631).jpg.
111. Jan Helebrant, https://commons.wikimedia.org/wiki/File:Chalcedony_quartz_SiO2_(26824918428).jpg.
112. James St. John, https://commons.wikimedia.org/wiki/File:Chalcedony_(Mojave_Desert,_California,_USA)_(33805717793).jpg.
113. Jan Helebrant, https://commons.wikimedia.org/wiki/File:Chalcedony_SiO2_(26017789432).jpg.
114. Jan Helebrant, https://commons.wikimedia.org/wiki/File:Chalcedony_quartz_SiO2_(39801423515).jpg.
117. James St. John, https://commons.wikimedia.org/wiki/File:Chalcedony-heulandite_(Apache_Creek,_Catron_County,_New_Mexico,_USA)_2_(33896524874).jpg.
120. JJ Harrison, https://commons.wikimedia.org/wiki/File:Botryoidal_Purple_Grape_Agate_Chalcedony_from_Indonesia.jpg.
124. Vassil, https://commons.wikimedia.org/wiki/File:British_Museum_Roman_Empire_18022019_Bust_of_a_young_man_Chalcedony_5810.jpg.
126. Tom Shearer photograph from *Agates Inside Out* (rights co-owned by Karen Brzys).
132. Tom Shearer photograph from *Agates Inside Out* (rights co-owned by Karen Brzys).
135. Tom Shearer photograph from *Agates Inside Out* (rights co-owned by Karen Brzys).
140. Tom Shearer photograph from *Agates Inside Out* (rights co-owned by Karen Brzys).
143. Tom Shearer photograph from *Agates Inside Out* (rights co-owned by Karen Brzys).
144. Tom Shearer photograph from *Agates Inside Out* (rights co-owned by Karen Brzys).
145. Tom Shearer photograph from *Agates Inside Out* (rights co-owned by Karen Brzys).
146. Tom Shearer photograph from *Agates Inside Out* (rights co-owned by Karen Brzys).
148. Shutterstock_747859927.
149. Recycled from a previous project. Source unknown.
156. Shutterstock_54294073.
157. Shutterstock_375133360.
158. Marie-Lan Nguyen, https://commons.wikimedia.org/wiki/File:Carnelian_intaglio_Prolemaic_queen_CdM_Paris.jpg.
159. James St. John, https://commons.wikimedia.org/wiki/File:Intergranular_calcite_in_autobrecciated_amygdaloidal_tholeiite_basalt_aa_lava_flow_top_(flow_B,_Two_Harbors_Basalts,_North_Shore_Volcanic_Series,_Mesoproterozoic,_1097-1098_Ma;_Burlington_Bay,_Two_Harbors,_Minnesota,_USA)_4.jpg.
160. James St. John, https://commons.wikimedia.org/wiki/File:Amygdaloidal_basalt_(Portage_Lake_Volcanic_Series,_upper_Mesoproterozoic,_1.093_to_1.097_Ga;_Keweenaw_Peninsula,_Upper_Peninsula_of_Michigan,_USA)_6.jpg.
161. Recycled from a previous project. Source unknown.
167. Jonathan Zander, https://commons.wikimedia.org/wiki/File:Native_Copper_Macro_Digon3-crop.jpg
168. UCL Mathematical and Physical Sciences, London, UK, https://commons.wikimedia.org/wiki/File:Native_copper_(12432818784).jpg.
169. Rajkiran Pericherla, https://commons.wikimedia.org/wiki/File:Copper_Lady.jpg.
170. James St. John, https://commons.wikimedia.org/wiki/File:Cupriferous_amygdaloidal_basalt_(Mesoproterozoic,_1.05-1.06_Ga;_Wolverine_Mine,_Kearsarge,_Upper_Peninsula_of_Michigan,_USA)_(17323753605).jpg.
171. James St. John, https://commons.wikimedia.org/wiki/File:Cupriferous_amygdaloidal_basalt_(Mesoproerozoic,_1.05-1.06_Ga;_Keweenaw_Peninsula,_northern_Michigan,_USA)_(16691110994).jpg.
172. James St. John, https://commons.wikimedia.org/wiki/File:Native_copper_in_conglomerate_(Keweenaw_Peninsula,_Michigan,_USA).jpg.
173. James St. John, https://commons.wikimedia.org/wiki/File:Cupriferous_amygdaloidal_basalt_(Kearsarge_Flow,_Portage_Lake_Volcanic_Series,_upper_Mesoproterozoic,_1.095_Ga;_South_Kearsarge_Mine,_Upper_Peninsula_of_Michigan,_USA)_1.jpg.
174. James St. John, https://commons.wikimedia.org/wiki/File:Copper_glacial_boulder_(Mesoproterozoic,_1.05-1.06_Ga;_Upper_Peninsula_of_Michigan,_USA)_(17145442128).jpg.
177. Recycled from a previous project. Source unknown.

180. James St. John, https://commons.wikimedia.org/wiki/File:Amygdaloidal_basalt_(Portage_Lake_Volcanic_Series,_upper_Mesoproterozoic,_1.093_to_1.097_Ga;_Keweenaw_Peninsula,_Upper_Peninsula_of_Michigan,_USA)_6.jpg.
181. James St. John, https://commons.wikimedia.org/wiki/File:Epidote_in_gneiss_(Precambrian;_Rt._93_roadcut_next_to_the_New_River,_Mouth_of_Wilson,_Virginia,_USA)_3_(30917533341).jpg.
185. James St. John, https://commons.wikimedia.org/wiki/File:Epidote_(Northeastern_University_Marine_Station,_Nahant_Peninsula,_Massachusetts,_USA)_1_(49741749621).jpg.
187. Strekeisen, https://commons.wikimedia.org/wiki/File:Epidote_explosion.JPG.
189. Tom Shearer photograph from *Agates Inside Out* (rights co-owned by Karen Brzys).
199. Shutterstock_175890668.
201. Shutterstock_353289287.
204. Didier Descouens, https://commons.wikimedia.org/wiki/File:Prehnite_Vall%C3%A9e_d%27Aure.jpg.

Chapter 4: Sedimentary Rocks
2. James St. John, https://commons.wikimedia.org/wiki/File:Bituminous_Coal_(Washington_Coal,_Upper_Pennsylvanian).jpg.
3. Gary Todd, https://commons.wikimedia.org/wiki/File:Fossil_Trilobite.jpg.
4. James St. John, https://commons.wikimedia.org/wiki/File:Folded_jaspilite_BIF_Hamersley_Range_Western_Australia.jpg.
5. Willem van Aken, CSIRO, https://commons.wikimedia.org/wiki/File:CSIRO_ScienceImage_4203_A_bluegreen_algae_species_Cylindrospermum_sp_under_magnification.jpg.
7. James St. John, https://commons.wikimedia.org/wiki/File:Jaspilite_banded_iron_formation_(BIF)_(Negaunee_Iron-Formation,_Paleoproterozoic,_1.874_or_2.11_Ga;_Jasper_Knob,_Ishpeming,_Michigan,_USA)_82_(47976336858).jpg.
8. Shutterstock_650426161.
11. GOKLuLe , https://commons.wikimedia.org/wiki/File:Magnetite_sample_with_neodymium_magnet.jpg.
12. Xvazquez, https://commons.wikimedia.org/wiki/File:Grypania_spiralis.JPG.
14. Shutterstock_137487155.
18. Woudloper, https://commons.wikimedia.org/wiki/File:Banded_Iron_Formation_Barberton.jpg.
19. James St. John, https://commons.wikimedia.org/wiki/File:Block_of_jaspilite_banded_iron_formation_(BIF)_in_wall_at_base_of_Jasper_Knob_(Ishpeming,_Michigan,_USA)_3_(48069775196).jpg.
20. Mike Beauregard, https://commons.wikimedia.org/wiki/File:A_banded_iron_formation.jpg.
21. James St. John, https://commons.wikimedia.org/wiki/File:Jaspilite_banded_iron_formation_(Soudan_Iron-Formation,_Neoarchean,_~2.69_Ga;_Rt._169_roadcut_between_Soudan_%26_Robinson,_Minnesota,_USA)_11_(18419358663).jpg.
22. James St. John, https://commons.wikimedia.org/wiki/File:Magnetite_banded_iron_formation_(Soudan_Iron-Formation,_Neoarchean,_~2.69_Ga;_Rt._169_roadcut_between_Soudan_%26_Robinson,_Minnesota,_USA)_21_(19035097072).jpg.
23. James St. John, https://commons.wikimedia.org/wiki/24. James St. John, https://commons.wikimedia.org/wiki/File:Claw-shaped_clast_in_basaltic_lapillistone_(Middle_Tholeiitic_Unit,_Kidd-Munro_Assemblage,_Neoarchean,_2.711-2.719_Ga;_just_east_of_the_Potter_Mine,_east_of_Timmins,_Ontario,_Canada)_1_(47076494094).jpg.
25. National Park Service, Cave Junction, USA, https://commons.wikimedia.org/wiki/File:Breccia_(9940637004).jpg.
26. James St. John, https://commons.wikimedia.org/wiki/File:Collapse_breccia_(Everton_Formation,_Middle_Ordovician;_Rush_Creek_District,_Arkansas,_USA)_2.jpg.
27. James St. John, https://commons.wikimedia.org/wiki/File:Fault_with_fault_breccia_in_interbedded_metagraywacke-slate_(Lake_Vermilion_Formation,_Neoarchean,_2.695-2.722_Ga;_Pike_River_Bridge_outcrop,_just_north_of_Peyla,_Minnesota,_USA)_2_(21622666821).jpg.
28. James St. John, https://commons.wikimedia.org/wiki/File:Volcanic_breccia.jpg.
29. James St. John, https://commons.wikimedia.org/wiki/File:Impact_breccia_(Eocene,_39_Ma;_Haughton_Impact_Structure,_Devon_Island,_northern_Canada)_3_(16677908618).jpg.
30. James St. John, https://commons.wikimedia.org/wiki/File:Azurite-cemented_breccia_(Phelps_Dodge_Morenci_Mine,_Arizona,_USA)_(14937041627).jpg.
32. Tommy from Arad, https://commons.wikimedia.org/wiki/File:Jasper_outcrop_in_the_Bucegi_Mountains.jpg.
33. Shutterstock_1021952959.
34. James St. John, https://commons.wikimedia.org/wiki/File:Basaltic_lapillistone_(Middle_Tholeiitic_Unit,_Kidd-Munro_Assemblage,_Neoarchean,_2.711-2.719_Ga;_just_east_of_the_Potter_Mine,_east_of_Timmins,_Ontario,_Canada)_17_(33979664758).jpg.

35. Tom Shearer photograph from *Agates Inside Out* (rights co-owned by Karen Brzys).
36. James St. John, https://commons.wikimedia.org/wiki/File:Limestone_breccia_or_brecciated_limestone_(Maxville_Limestone,_Middle_to_Upper_Mississippian;_likely_derived_from_a_quarry_in_southwestern_Muskingum_County,_Ohio,_USA)_1_(32394790973).jpg.
37. Shutterstock_75814477.
38. Geologicharka, https://commons.wikimedia.org/wiki/File:Tectonic_breccia.JPG.
40. Shuterstock_787838053.
44. James St. John, https://commons.wikimedia.org/wiki/File:Autobrecciated_amygdaloidal_quartz_tholeiite_basalt_aa_lava_flow_top_(flow_B,_Two_Harbors_Basalts,_North_Shore_Volcanic_Series,_Mesoproterozoic,_1097-1098_Ma;_Burlington_Bay,_Two_Harbors,_Minnesota,_USA)_31_(22290589240).jpg.
48. James St. John, https://commons.wikimedia.org/wiki/File:Fossiliferous_chert_(Upper_Mercer_Flint,_Middle_Pennsylvanian;_Nellie_West_Outcrop,_Coshocton_County,_Ohio,_USA)_1_(31882402546).jpg.
50. Gordon T. Taylor, https://commons.wikimedia.org/wiki/File:Diatoms_through_the_microscope.jpg.
51. Shutterstock_441244423.
58. James St. John, https://commons.wikimedia.org/wiki/File:Crinoidal_vuggy_chert_(Carboniferous;_limestone_quarry_near_Komsomolske,_southeastern_Ukraine)_-_1.jpg.
65. St. John, https://commons.wikimedia.org/wiki/File:Chert_nodule_(%22Indiana_hornstone%22)_(probably_Mississippian;_Indiana,_USA)_6_(45494087962).jpg.
66. Shutterstock_369686234.
68. James St. John, https://commons.wikimedia.org/wiki/File:Jaspilite_banded_iron_formation_(BIF)_(Temagami_Iron-Formation,_Neoarchean,_2.736_Ga;_Sherman_Iron_Mine,_Temagami,_Ontario,_Canada)_4.jpg.
71. James St. John, https://commons.wikimedia.org/wiki/File:Araucarioxylon_arizonicum_(fossil_wood)_(Chinle_Formation,_Upper_Triassic;_south_of_Adamana,_Arizona,_USA)_1_(26667489168).jpg.
77. 86. James St. John, https://commons.wikimedia.org/wiki/File:Jasper_(32132824820).jpg.
87. ZeWrestler, https://commons.wikimedia.org/wiki/File:Conglomerate_Rock.jpg.
89. NASA/GSFC/METI/ERSDAC/JAROS, and U.S./Japan ASTER Science Team, https://commons.wikimedia.org/wiki/File:Alluvial_fan,_Taklimakan_Desert,_XinJiang_Province,_China,_NASA,_ASTER.jpg.
100. James St. John, https://commons.wikimedia.org/wiki/File:Jasper-quartz_pebble_conglomerate_(Lorrain_Formation,_Paleoproterozoic,_~2.3_Ga;_Ottertail_Lake_Northeast_roadcut,_near_Bruce_Mines,_Ontario,_Canada)_38_(32766102277).jpg.
101. Slim Sepp, https://commons.wikimedia.org/wiki/File:00142_9_cm_conglomerate.jpg.
103. Alexander Bliss, https://commons.wikimedia.org/wiki/File:QUERN_(FindID_880789).jpg.
104. James St. John, https://commons.m.wikimedia.org/wiki/File:Quartz-pebble_conglomerate_(Sharon_Conglomerate,_Lower_Pennsylvanian;_Toboso_East_railroad_cut,_Licking_County,_Ohio,_USA)_2_(34394301115).jpg.
105. James St. John, https://commons.wikimedia.org/wiki/File:Quartz-pebble_conglomerate_(%22Sharon_Conglomerate%22,_Lower_Pennsylvanian;_Jackson_North_roadcut,_Ohio,_USA)_43_(26244648937).jpg.
106. James St. John, https://commons.wikimedia.org/wiki/File:Polymict_conglomerate_(Ogishkemuncie_Conglomerate,_Neoarchean;_Ogishkemuncie_Lake,_Boundary_Waters,_Minnesota,_USA)_2_(22133918728).jpg.
107. Shutterstock_254156944.
108. James St. John, https://commons.wikimedia.org/wiki/File:Quartz-pebble_conglomerate_(Sharon_Conglomerate,_Lower_Pennsylvanian;_Toboso_East_outcrop,_Licking_County,_Ohio,_USA)_2.jpg.
110. Eurico Zimbres, https://commons.wikimedia.org/wiki/File:Metaconglomerate.jpg.
111. Amytrippmyers, https://commons.wikimedia.org/wiki/File:Crestone_Conglomerate_in_Colorado_USA.jpg.
113. Karen Brzys diagram with information from https://commons.wikimedia.org/wiki/File:Ft_Pore.png.
115. http://www.earth.ox.ac.uk/~oesis/micro/sedimentary/index.html#img=greywacke_pm19-16.jpg.
116. Shutterstock_1020531019.
117. James St. John, File:Graywacke (16176003404).jpg - Wikimedia Commons.
118. Deamstime_xxl_132277560.
120. James St. John, https://commons.wikimedia.org/wiki/File:Interbedded_graywacke-siltstone-slate_(Mud_Lake_sequence,_Neoarchean;_Bourgin_Road_roadcut,_Virginia,_Minnesota,_USA)_2_(22878926504).jpg.

229

121. James St. John, https://commons.wikimedia.org/wiki/File:Glacial_striations_from_Pleistocene_glaciation_on_Mud_Lake_sequence_sedimentary_rocks_(Neoarchean;_Bourgin_Road_roadcut,_Virginia,_Minnesota,_USA)_12_(23416542261).jpg.
124. James St. John, https://commons.wikimedia.org/wiki/File:Concretion_in_the_Dakota_Sandstone_(Lower_Cretaceous)_(Dinosaur_Ridge,_Colorado,_USA).jpg.
125. James St. John, https://commons.wikimedia.org/wiki/File:Concretion_(Vinton_Member,_Logan_Formation,_Lower_Mississippian;_Hanover_Pit,_Licking_County,_Ohio,_USA)_2_(32587163907).jpg.
141. Modified from Monazite1982, https://commons.wikimedia.org/wiki/File:Glauconite-quartz_sandstone_(foundation_of_the_Desyatynna_Church_in_Kyiv).jpg.
142. Chris857, https://commons.wikimedia.org/wiki/File:Jacobsville_Sandstone_sample_2.jpg.
144. Chris857, https://commons.wikimedia.org/wiki/File:Pallet_of_Jacobsville_Sandstone_blocks.jpg.
165. Shutterstock_208168519.
167. https://en.wikipedia-on-ipfs.org/wiki/Ooid.html.
172. P.Cikovac, https://commons.wikimedia.org/wiki/File:Identification_of_limestone_(carboniferous_from_the_High_Karst_nappe)_with_hydrochloric_acid.jpg.
177. Shutterstock_56061232.
179. Recycled from a previous project. Source unknown.
181. James St. John, https://commons.wikimedia.org/wiki/File:Limestone_(Sisson_Member,_St._Louis_Limestone,_Middle_Mississippian;_Putnam_County,_Indiana,_USA)_(34806273141).jpg.
185. James St. John, https://commons.wikimedia.org/wiki/File:Oolitic_limestone_(Salem_Limestone,_Middle_Mississippian;_Bedford,_Indiana,_USA)_2.jpg.
186. Shutterstock_292743518.
187. Shutterstock_2011877.
188. Shutterstock_92689663.
189. James St. John, https://commons.wikimedia.org/wiki/File:Neuropteris_flexuosa_fossil_plant_(Mazon_Creek_Lagerstatte,_Francis_Creek_Shale,_Middle_Pennsylvanian;_coal_mine_dump_pile_near_Essex,_northern_Illinois,_USA)_(14910119354).jpg.
190. James St. John, https://commons.wikimedia.org/wiki/File:Black_shale_(New_Albany_Shale,_Upper_Devonian;_Clark_County,_Indiana,_USA)_(41694033411).jpg.
191. James St. James, https://commons.wikimedia.org/wiki/File:Potholes_(Presque_Isle_River,_Porcupine_Mountains_State_Park,_Upper_Peninsula_of_Michigan,_USA)_(21466113862).jpg.
192. Shutterstock_111451562.
195. Antandrus,, https://commons.wikimedia.org/wiki/File:JuncalShaleWeathered.jpg.
197. Shutterstock_116769319.
201. James St. James, https://commons.wikimedia.org/wiki/File:Cladoselache_fyleri_fossil.jpg.
202. Recycled from a previous project. Source unknown.
205. James St. James, https://commons.wikimedia.org/wiki/File:Intensely-burrowed_shale_(Pottsville_Group,_Middle_Pennsylvanian;_Rock_Cut,_Muskingum_County,_Ohio,_USA)_4_(36593089334).jpg.
206. Shutterstock_356552168.
207. James St. John, https://commons.wikimedia.org/wiki/File:Weathered_shale_chips_(Sunbury_Shale,_Lower_Mississippian;_Bentleyville,_Ohio,_USA)_(34897841386).jpg.
208. James St. John, https://commons.wikimedia.org/wiki/File:Waldron_Shale_(Middle_Silurian)_(St._Paul_Stone_Quarry,_St._Paul,_Indiana,_USA)_4_(21833021510).jpg.
209. Shutterstock_376750213.
210. Shutterstock_1032792517.

Chapter 5: Metamorphic Rocks
4. USGS, https://www.usgs.gov/media/images/garnet-schist.
5. Kmtextor, https://commons.wikimedia.org/wiki/File:Slatewall_trebarwith_strand.jpg.
7. James St. John, https://commons.wikimedia.org/wiki/File:Marble_(Yule_Marble,_Middle_Miocene,_12_Ma;_Marble,_northern_Gunnison_County,_western_Colorado_USA)_2_(16887651452).jpg.
8. James St. John, https://commons.wikimedia.org/wiki/File:Quartzite_(Precambrian;_Rock_Creek_Canyon,_Beartooth_Mountains,_Montana,_USA)_1.jpg.
9. Ildar Sagdejev, https://commons.wikimedia.org/wiki/File:2008-06-26_Stacked_roofing_slate_1.jpg.
11. James St. John, https://commons.wikimedia.org/wiki/File:Phyllite_(French_Slate,_Paleoproterozoic;_Snowy_Range_Road_roadcut,_Medicine_Bow_Mountains,_Wyoming,_USA)_5_(45574603622).jpg.
12. Shutterstock_1008857698.
15. Shutterstock_746532748.

18. James St. James, https://commons.wikimedia.org/wiki/File:Furcaster_paleozoicus_St%C3%BCrtz,_1886_ (8.5_cm_long) _with_current_aligned_arms_from_the_Hunsr%C3%BCck_Lagerst%C3%A4tte_ (Lower_Devonian)_of_western_Germany._(8474368510).jpg.
21. Shutterstock_12045859.
23. Ashley Columbus, https://commons.wikimedia.org/wiki/File:Nantlle_slate.JPG.
24. Steve Snodgrass, https://commons.wikimedia.org/wiki/File:Pieces_o_Slate_(3677978933).jpg.
25. James St. John, https://commons.wikimedia.org/wiki/File:Slate_(Knife_Lake_Formation,_metamorphism_at_2.7_Ga,_Neoarchean;_Rt._135_roadcut,_Gilbert,_Minnesota,_USA)_5_(23399454252).jpg.
26. Shutterstock_1039729399.
28. Andreas F. Borchert, https://commons.wikimedia.org/wiki/File:Gleann_Cholm_Cille_Turas_Cholmcille_Stad_5_Central_Cairn_Slab_East_Face_2010_09_24.jpg.
29. Slim Sepp, https://commons.wikimedia.org/wiki/File:Glaucophane_schist_ss_0,5_xp_2007.jpg.
31. Zarmel, https://commons.wikimedia.org/wiki/File:Schist_and_Quartzite_in_Brittany_France.jpg.
34. James St. John, https://commons.wikimedia.org/wiki/File:Kyanite_schist_1_(31284731697).jpg.
35. James St. John, https://commons.wikimedia.org/wiki/File:Garnet_schist_1_(16735443408).jpg.
36. James St. John, https://commons.wikimedia.org/wiki/File:Muscovite_schist_(Appalachian_Gap,_Green_Mountains,_Vermont,_USA)_5.jpg.
37. Michael C. Rygel, https://commons.wikimedia.org/wiki/File:Schist_detail.jpg.
41. Chd, https://commons.wikimedia.org/wiki/File:Garnet-mica-shist-x-nicols.jpg.
44. James St. John, https://commons.wikimedia.org/wiki/File:Garnet-chlorite_schist_(Ducktown,_Tennessee,_USA)_(27022679207).jpg.
45. James St. John, https://commons.wikimedia.org/wiki/File:Muscovite_schist_(Paleoproterozoic;_Black_Hills,_South_Dakota,_USA)_2_(31511141474).jpg.
46. James St. John, https://commons.wikimedia.org/wiki/File:Chlorite_schist_(Wissahickon_Schist,_Neoproterozoic_to_Cambrian;_Jarrettsville,_Maryland,_USA)_(16735383710).jpg.
47. Amcyrus2012, https://commons.wikimedia.org/wiki/File:Mica_shist_student_sample.JPG.
48. James St. John, https://commons.wikimedia.org/wiki/File:Talc_schist_1_(16922837125).jpg.
49. James St. John, https://commons.wikimedia.org/wiki/File:Tourmaline-mica_schist_(near_Custer,_South_Dakota,_USA)_(40158163570).jpg.
50. Amcyrus2012, File:Mica Shist2.JPG - Wikimedia Commons.
51. James St. John, https://commons.wikimedia.org/wiki/File:Chlorite_schist_(Wissahickon_Schist,_Neoproterozoic_to_Cambrian;_Jarrettsville,_Maryland,_USA)_(16735383710).jpg.
52. Amcyrus2012, File:Mica Shist2.JPG - Wikimedia Commons.
55. Mike Beauregard, https://commons.wikimedia.org/wiki/File:4,030,000,000_Years_Acasta_Gneiss.jpg.
56. James St. John, https://commons.wikimedia.org/wiki/File:Garnet_paragneiss_Nuvvuagittuq_Greenstone_Belt,_4.28_Ga.jpg.
62. James St. John, https://commons.wikimedia.org/wiki/File:Gneiss_(Archean;_Ennis_Lake_North_roadcut,_Madison_County,_Montana,_USA)_1_(44800968564).jpg.
66. Shutterstock_86652253.
69. Michael C. Rygel, https://commons.wikimedia.org/wiki/File:Gneiss_detail.jpg.
76. https://www.si.edu/newsdesk/factsheets/national-museum-natural-history.
78. James St. John, https://commons.wikimedia.org/wiki/File:Potassium_feldspar_(32499528651).jpg.
80. Shutterstock_271751006.
88. Shutterstock_206934004.

Appendix G: Index

This index was compiled by using the "automatic" idexing function in InDesign. Although it is better to have this function than to not, it does not work perfectly. I had to separately compile an index for each chapter, cut and paste each into Microsoft Word, edit and sort the index listing, and then cut and paste the compiled index here. Unfortunately, the InDesign indexing function requires separate searches for the same word with and without capitalization. I did my best to do be comprehensive in the indexing, but I may have missed flagging some entries.

The index has been organized into sections: Geology Terms and History, Minerals, Rocks, United States Specimen Photos, and International Specimen Photos.

Geology Terms and History

Alluvial fan: 148, 150
Amygdaloidal: 67-68, 75-78, 80-81, 83, 93, 107, 111, 113, 116, 119
Amygdules: 75
Basalt columns: 72
Batholith: 16-17, 19-20, 38
Black River Pathway Park (Michigan): 149
Botryoidal: 94, 96
Cave of the Mounds (Wisconsin): 174
Cementation: 120, 147, 156
Circumstellar disk: 9
Clast supported: 130, 134-135
Coarse-grained: 19, 21, 24, 26, 39, 46, 49, 54, 70, 123, 154, 189, 195, 208
Colloids: 76
Compaction (sedimentary): 120-121, 147, 156, 172, 179
Conchoidal fracture: 94, 139, 140, 142
Continental crust: 64
Continental plate: 22, 39, 47
Continental rift zone: 54
Convection currents: 23, 61
Convergent plate boundary: 39
Core-mantle boundary: 63
Craton: 15, 60
Cross-polarized microscopic image: 6, 18, 34, 38, 41, 48, 58, 70, 133, 196, 200
Cryptocrystalline: 6
Crystal: 5, 10, 14, 17-18
Crystallization 18, 32, 34, 47, 54
Cyanobacteria: 13-14, 124-125
Deposition (sedimentary): 120-121, 137, 147, 155-156, 163
Diatoms: 138
Dikes: 16-17, 19, 39
Drusy (crystals): 94, 115

Earth's core: 10, 12
Earth's crust: 2, 9, 11-14, 64-65, 71
Earth's mantle: 60-61
Earthquakes: 64
Electron microscope image: 25
Erosion (erosional forces): 24, 120, 131, 142, 154
Evaporites: 121
Extrusive Igneous Rocks: 7, 60-119
Fibrous crystals: 98
Fine-grained (rock texture): 67, 70, 74, 78-79, 82, 85-87, 112-123, 147, 181, 189, 191
First chalcedony layer: 97-98
Fluorescent (fluoresce): 8, 21, 54-59, 93- 94, 96, 176
Foliated (metamorphic): 187-190, 197-198, 202-203, 205, 207
Geode: 94, 98
Geologic Background: 2-4, 9, 15, 60, 120, 185
Glacial till: 69, 104
Glaciers: 70, 98, 104, 120, 150, 158-159
Grand Canyon: 155
Great Lakes: 2, 3, 65, 67, 71, 92, 123, 159, 173
Grenville Provence: 62
Groundmass: 67, 74, 78, 82-83, 85-87
Grypania spiralis: 126-127
Hadean Eon: 10, 11, 13
Hot spot mantle plume: 63-64
Huron Mountains: 148
Hydrothermal activity (fluids): 15, 18, 35, 76-77, 113, 189, 208
Impact breccia: 132
in situ: 109
Indigenous people: 110, 138
Inflow channel (agate): 101
International Mineralogical Society: 14
Iron meteorites: 9
Island arcs: 69, 82
Isotopes: 10-12
Jasper Knob (Ishpeming, Michigan): 125-126
Keweenaw Peninsula: 66, 107-108, 110, 112, 118-119,148
Kinetic energy: 17- 8
Lake Huron: 150-151
Lake Superior beach (rock): 30, 36, 44, 49, 51, 53, 55, 59, 78, 80, 85, 89-90, 114, 117, 200, 206-207, 211
Late Heavy Bombardment: 11
Laurentia (plateau/continent): 60, 62, 173
Lava: 15, 16, 47, 65, 66, 67, 68, 69, 70, 71, 72, 73, 75, 77, 79, 82, 83, 85, 86, 88, 92, 136
Lithification: 121, 151, 155-156, 179
Lower Peninsula: 150
Mackinac Island: 178

Macrocrystalline: 5-7, 93, 94, 115
Magma chamber: 68
Magma: 15-19, 27-28, 32-33, 35, 38-40, 43, 46-47, 51, 53-54, 63-64, 68-69, 72, 74, 82, 85, 86, 88
Mantle plume: 63-64
Mars: 63, 69
Matrix supported (breccia): 130, 134-136
Medium-grained (rock texture): 19, 195
Mesabi Range (Minnesota): 148
Metamorphism, burial: 185
Metamorphism, contact: 185
Metamorphism, regional: 185
Methane: 124
Microcrystalline: 6
Midcontinent Rift: 63-68, 70-71, 76, 82, 110, 112, 144, 148, 161-162
Mid-ocean ridges: 47, 69
Mohs Hardness Scale: 5, 27, 35, 42, 50, 87, 93, 112, 115, 118, 119, 192, 206, 210, 127, 134, 137, 140, 144, 152, 158, 165, 173, 176, 179, 182
Moon: 11, 12, 13
Mount Everest: 71
Nanometers: 55
Napoleon Bonaparte: 105
North American continent: 60-61, 65
Northern Michigan Highlands: 161
Number of minerals (on Earth): 9, 11, 13-14
Ocean crust: 6
Ocean plates: 39
Olympus Mons (Mars volcano): 69
Orbicular 38, 45, 86, 88, 90
Oxygen (source): 10, 11, 14
Phenocrysts: 67, 74, 79, 82-83, 86-87, 89, 91
Phosphorescent: 96
Photosynthetic: 124
Phytoplankton: 120
Pictured Rocks National Lakeshore: 123, 168, 171
Piezoelectricity: 138
Pit-marked (agate husk): 100-102
Plate boundaries: 186, 190
Plate tectonic (forces): 17, 19, 22, 23, 38, 39, 47, 125, 126, 129, 155, 158, 173, 185, 190, 194, 197, 202, 205, 206, 209
Porcupine Mountain (Michigan State Park): 148-149, 180, 183-184
Porphyritic: 50, 67, 74, 78, 81-82, 85, 90
Precambrian time: 60
Pretty rock syndrome: 1
Pseudomorph: 93
Pyroclastic (volcanic): 67-68, 85-86, 132

Radioactive elements: 63
Radiolarians: 138
Rift valley (rift): 64-66, 69-71, 76, 82
Rodinia (supercontinent): 62, 63
Sagenite, (agate): 97, 103
Scanning electron microscope (photo): 6
Secondary fill: 81, 114, 116, 143
Sediments, biologic: 120
Sediments, chemical: 121
Sediments, clastic: 120
Sediments, cosmogenous: 121
Sills: 16, 19
Solar System: 9-13, 69
Split Rock Lighthouse (Minnesota): 72
Stalactic habit (chalcedony): 94
Statue of Liberty: 109
Subduction zones: 38, 39, 47, 54, 190
Sun: 9-10, 100
Superior Craton: 60, 62
Supernova: 9, 10
Syncline: 65-66, 71
Tectonic plates: 22-23, 122, 186
Theia: 12-13
Ultraviolet light (UV): 55
Upper Peninsula: 108, 110, 126, 127, 162, 171, 176, 180, 193
Vesicle pockets: 67, 71-73, 75-78, 85-86, 88-89, 92-93, 97, 112-113
Vesicular: 67, 72-73, 78, 80, 88
Vinegar: 140, 173, 175
Viscosity (viscous): 32-33
Volcanic ash: 132
Volcano: 13, 69, 85, 92
Waxy luster: 102, 139-140, 142, 146
Weathering: 19, 24-25, 120, 131, 142, 155, 174, 179
Wishing stone: 116
Zircons: 10
Zooplankton: 120

Minerals
(Agate) water-level banding: 103
(Agate), sagenite: 97, 103
Agate eye formations: 103
Agate: 1, 6, 68, 97-99, 75-77, 81, 86, 93, 97-106
Amethyst: 6
Amphibole: 41
Amygdaloidal Minerals: 8, 68
Apatite: 32
Aquamarine: 32
Beryl: 32, 86

Calcite: 68, 75, 78, 107-108, 118, 121-122, 130, 154, 155, 156, 163, 174, 178, 179-180, 187
Calcium carbonate 123, 138, 147, 151, 172-173
Carnelian: 68, 75, 93, 104-106
Chalcedony rose: 94-95
Chalcedony: 68, 75, 93-96
Chlorastrolite: 68, 75, 108-109
Chlorite: 156, 187, 190, 195, 200-201
Chrysoprase: 93
Clay/clay minerals: 24, 120-123, 147, 151, 155-156, 159, 163, 179
Copper: 68, 75, 109-112, 118
Corundum (ruby): 32
Datolite: 68, 75, 112, 118
Emerald: 32
Epidote: 44, 46, 68, 75, 78, 80, 107, 113-114, 156-157, 187, 208-209, 212
Feldspar: 5, 24, 25, 27-28, 30, 32-34, 36, 38-39, 41, 43, 46, 49-52, 54, 69, 154-156, 159, 162, 187, 195, 202, 208-210, 212
Fluorite: 32
Garnet: 33, 89, 187
Gold: 117
Gypsum: 70, 121, 155
Hackmanite: 55
Hematite: 121, 124, 126, 150, 155
Hornblende: 5, 28, 30, 68-69, 82, 86, 187, 195, 202, 207
Iron: 51, 121, 123-128, 130, 137-138, 143, 144, 147, 151, 156, 163-164, 173, 177, 179, 180
Isle Royale Greenstone: 68, 75, 108
Jasper: 81, 86, 93, 123-126, 128, 133, 137, 143-146, 150-151, 209-210
Kyanite: 187, 198
Limonite: 99, 101, 102, 121, 128, 137, 145-146, 155
Magnetite: 38, 41, 124, 126
Mica: 5, 24-25, 27-28, 32, 34, 36, 38-39, 43, 54, 156, 179, 187, 190, 195-196, 197-198, 200-202, 207
Milky quartz: 68, 75, 115-117
Minerals (definition): 4
Moganite: 93
Olivine: 46, 51-52, 68-69, 74, 82, 187
Onyx: 93
Opal: 86, 93
Patricianite: 118
Prehnite: 68, 75, 118-119
Pumpellyite: 108
Pyroxene: 41, 46, 50, 68-69, 74, 82
Quartz: 4-5, 24, 27-28, 49, 68-69, 75, 78, 81, 93-94, 107, 113, 115-117, 137-138, 144-145, 150, 154-157, 159, 179, 187-188, 195, 197, 199-200, 202, 207, 209, 212
Silica: 121-124, 126, 129-130, 137-138, 142-147,
Sillimanite: 187

Sodalite: 54-55, 57
Staurolite: 187
Talc: 201
Thomsonite: 68, 75, 118-119
Tin: 32
Topaz: 32, 86
Tourmaline: 32-33, 201
Tungsten: 32

Rocks
Acasta gneiss: 203
Amygdaloidal Basalt: 67, 75, 78, 80-81, 111, 116
Andesite: 40, 68, 74, 82-84
Banded Iron Formation (BIF): 123-129
Basalt: 15, 17, 21, 31, 40, 46, 47, 67-82, 85, 88, 93, 97-99, 105, 107-108, 110-113, 116, 118-119, 131, 136, 143, 156, 157, 159
Breccia, sedimentary: 131
Breccia, tectonic: 131
Breccia, volcanic: 132
Breccia: 8, 123, 130-136, 151
Chert: 86, 93, 123-124, 126, 131, 137-144, 150, 156
Coal: 120-122
Concretion: 93, 159-160
Conglomerate: 8, 123, 130, 133, 147-153, 168
Copper Harbor Conglomerate: 123, 148-149
Diabase: 21, 46
Diorite: 21, 39, 41, 42, 43, 44
Flint: 86, 93, 137-138
Gabbro: 21, 31, 39, 43, 46-53, 70, 74
Gneiss: 30, 156, 162, 187-189, 191, 196, 198, 199, 202-208, 210
Granite: 5, 15, 20-32, 34, 36-41, 43, 46, 49, 51, 54, 59, 131, 162, 189, 202, 203, 208-210
Granitoid rocks: 23
Granodiorite: 21, 38
Graywacke: 79, 123, 156-159
Intrusive Igneous Rocks: 7, 15-59
Jacobsville Sandstone: 123, 161-169
Jaspilite: 124, 144
Limestone: 120, 123,131, 135, 138, 140, 172-178
Septarian nodule: 136
Marble: 187-188
Metamorphic Rocks: 8, 185-212
Munising Formation Sandstone: 123, 168-172
Nepheline Syenite: 59
Nonesuch Shale: 180, 183-184
Obsidian: 138
Omarolluk (Omar): 123, 159-160

237

Pegmatite: 5, 21, 29, 33-34, 35-36
Petrified wood: 143
Phyllite: 190-191
Porphyritic Basalt: 67, 74, 78, 81
Puddingstone: 23, 150-151
Pumice: 68, 73, 86, 92
Quartzite: 150, 156, 188, 197
Rhyolite: 68, 73-76, 82, 85-91, 97
Rock Identification: 1, 2, 7, 26, 34, 41, 48, 57, 78, 86, 93, 126, 133, 139, 144, 151, 157, 164, 169, 175, 181, 191, 196, 203-204, 209
Rocks (definition): 2, 5
Sandstone: 123, 131, 154-156, 159-166, 168-172, 188
Schist: 189, 195-200
Schorl: 33
Scoria: 67, 73
Sedimentary Rocks: 8, 120-185
Shale: 123, 156, 158, 179-184
Slate: 156, 182, 187, 189, 191-193, 195
Stromatoporaid: 177
Syenite: 21, 54, 57-59
Travertine: 173
Trilobite fossil: 122, 178
Unakite: 113-114,189, 208-212
Wonderstone (rhyolite): 91
Yooperlite: 55

United States Specimen Photos (35 states)
Note: This list includes some states for which geologic information has been included.
Alabama: 65
Alaska: 17
Arizona: 92, 110, 117, 133, 146, 155
Arkansas: 119, 131
California: 17, 20, 22, 38, 88, 92, 95
Colorado: 29, 36, 90, 119, 153, 160
Hawaii: 72
Idaho: 92
Indiana: 65, 142, 178, 184
Iowa: 65
Kansas: 65, 92
Kentucky: 65, 150
Maine: 27
Maryland: 201
Michigan: 64, 106, 107, 108, 110, 112, 113, 118-119, 123, 125-127, 129, 143, 149-150, 161-162, 168, 171, 176, 178, 180, 183-184, 93
Minnesota: 31, 53, 65, 107, 119, 129, 136, 153, 158, 194
Montana: 17
Nebraska: 65

Nevada: 83, 92
New Jersey: 119
New Mexico: 92
North Carolina: 33
Ohio: 65, 135, 150, 152-153, 183
Oklahoma: 65
Oregon: 92, 119, 130
Pennsylvania: 86
South Dakota: 20, 36, 94, 95-96, 200
Tennessee: 65, 200
Utah: 73, 74, 91
Vermont: 198
Virginia: 113
Washington, D.C.: 208
Wisconsin: 36, 59, 65, 174
Wyoming: 84, 90
Massachusetts: 54

International Specimen Photos (18 other countries)
Afghanistan: 55
Africa: 105
Antarctica: 45, 53
Australia: 10-11, 36, 38, 124
Brazil: 36, 97-98, 153
Canada: 55, 129, 132, 150, 159, 203-204
    Belcher Islands: 159
    Ontario: 130, 150, 211
    Quebec: 30
China (Chinese): 97-98, 104, 124
Egypt (Egyptian): 40, 105
England: 157
German: 193
Indonesia: 96
Norway: 34, 104
Romania: 133, 152
Russia: 53
Scotland (Scottish): 30, 48
Slovakia: 84
South Africa: 52, 128
Sweden: 53, 104

List of Rocks and Minerals Collected

List the rocks and minerals found in the left colmn below and the date and loication found in the right column.
